14.60

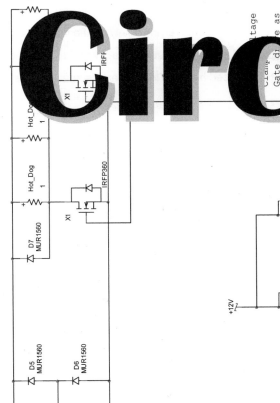

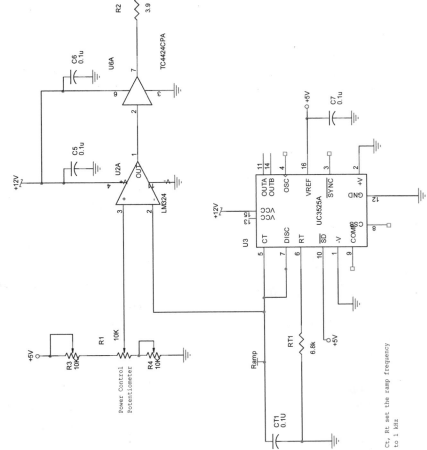

Cool Circuits

Marc E. Herniter

Associate Professor
ECE Department
Rose-Hulman Institute of Technology

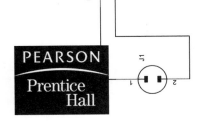

PEARSON

Prentice Hall

Upper Saddle River, New Jersey
Columbus, Ohio

NOTICE TO THE READER

Acquisitions Editor: Kate Linsner
Production Editor: Rex Davidson
Design Coordinator: Diane Ernsberger
Editorial Assistant: Lara Dimmick
Cover Designer: Ruta Fiorino/ fruitiDesign
Cover art: Index Stock
Production Manager: Matt Ottenweller
Marketing Manager: Ben Leonard

This book was printed and bound by Banta Book Group. The cover was printed by The Lehigh Press, Inc.

Pearson Education Ltd.
Pearson Education Singapore Pte. Ltd.
Pearson Education Canada, Ltd.
Pearson Education—Japan

Pearson Education Australia Pty. Limited
Pearson Education North Asia Ltd.
Pearson Educación de Mexico, S.A. de C.V.
Pearson Education Malaysia Pte. Ltd.

10 9 8 7 6 5 4 3 2

ISBN: 0-13-119343-0

Trademarks

"Multisim" and "Electronics Workbench" are registered trademarks of Interactive Image Technologies Limited.

"Microsoft," "Win32s," "MS-DOS," and "Windows" are registered trademarks of Microsoft Corporation.

MicroSim®, Orcad®, Orcad Capture®, Orcad Capture CIS®, Orcad PSpice®, Orcad PSpice A/D®, Probe®, PSpice®, and Schematics® are registered trademarks of Cadence Design Systems, Inc.

"Klingon" is a trademark of Paramount Pictures.

"Star Trek: The Next Generation" is a trademark of Paramount Pictures.

Preface

This book attempts to answer the common questions, "Why are we doing this?" and "What is this used for?" that students ask in the first and second analog electronics courses taught at most institutions. In a classic field like analog electronics it is hard to generate much student enthusiasm. Our culture has trained people to think that digital is the way of the future. Computers are digital. Cell phones are digital. Cameras are digital. TV is becoming digital. Everything good one day will be digital. Students do not see where analog electronics fit into their lives, nor do they see any cool applications of analog electronics. This is about to change.

This book discusses several demonstrations and design examples with the express purpose of showing students some of the cool things that can be done with analog electronics. These examples encourage students to take higher-level analog electronics courses and show students where analog electronics fit into the overall engineering picture. The examples are targeted at the first and second electronics courses in a two-course sequence. Some projects for the first course are a birthday candle blower that uses a diode as a temperature sensor, a bug sucker that uses an infrared emitter/detector pair to sense the presence of an insect, a battery-operated 18 V to 1000 V boost converter stun gun, a 24 V to 600 V voltage multiplier that looks like a Klingon pain stick, and a DC motor speed controller.

The purpose of the text is to generate engaging examples that make students more interested in analog electronics than in any other course, as well as to illustrate some of the basic principles covered in the class. For example, the bug sucker shows the use of a BJT as a switch; the stun gun illustrates the principle that the current through an inductor cannot go to zero instantaneously; and the BJT thermal instability demonstration illustrates the negative temperature coefficient of BJTs. Each example is a complete circuit design, and most of the circuits are at a level where students can understand the function of every component in the circuit, so students can understand, design, and build circuits that perform non-trivial functions at an early stage in their engineering careers.

The examples can be used as supplementary reading to inspire students or to create classroom demonstrations.

Comments and Suggestions

The author would appreciate any comments or suggestions on this text. Suggestions for new cool circuits are especially welcome. Please feel free to contact the author using any of the methods listed below:

- **E-mail:** Marc.Herniter@ieee.org
- **Phone:** (812) 877-8512
- **Fax:** (812) 877-8895
- **Mail:** Rose-Hulman Institute of Technology, CM123, 5500 Wabash Avenue, Terre Haute, IN 47803-3999

Acknowledgments

I would like to thank my students at Rose-Hulman Institute of Technology for giving me the inspiration to create this manual. They made it clear to me that this book is necessary, and they rewarded me with interest and enthusiasm. I would also like to thank Patrick Worland of Raytheon Missile Systems and James C. Brietkrietz of Honeywell Aerospace Electronic Systems for taking the time to read this book and not say anything particularly bad about it. In fact, they made some cool comments that really helped to improve this book. They are living proof that if you read this book, you too could one day do cool stuff in analog electronics.

Contents

Cool Circuit I
Birthday Candle Blower

Our first cool circuit is a birthday candle blower that repeatedly blows out one of those annoying trick birthday candles that relights itself. We have all been to a birthday party where the host secretly uses candles that magically relight themselves. The unsuspecting guest of honor then proudly blows out all of the candles only to be "surprised" that the flames come back to life. The person then blows them out several more times and proceeds to pass out, start hyperventilating, or spit all over the cake. These candles are never really a surprise because the candles have a distinctive flame that sparks, but they are fun nonetheless. If you build our birthday candle blower, you never need to blow out these candles again. The next time you are the guest of honor at a birthday party, you can bring along this circuit and have the last laugh. When you notice the magic candles, you can whip out your candle blower and be the life of the party.

I.A. Temperature Sensor

This circuit uses a diode as a temperature sensor to sense the flame of a candle. When the diode becomes warm enough, a fan turns on and blows out the candle. The diode then cools down and the fan turns off. The candle then magically relights itself and heats up the diode again. The diode senses the rise in temperature and turns on the fan again, which blows out the candle again. This oscillating feedback loop continues until the candle stops relighting itself or the candle is used up.

This circuit illustrates the use of a semiconductor PN diode as a temperature sensor, a DC power supply, and a temperature sensor to control a cooling fan. Recall that if the current through a PN diode is held constant, its forward voltage will decrease by approximately 2 mV for every 1°C rise in the PN junction temperature. [1] We will illustrate this concept by simulating the current-source-based temperature sensing circuit below.

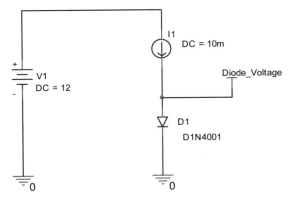

The constant current source (I_1) keeps the current through the diode constant at 10 mA. We then set up a temperature sweep to vary the temperature from 15°C to 40°C. Since the current is held constant by the current source, only the diode voltage can vary as the temperature is changed. The results are shown in the following graph.

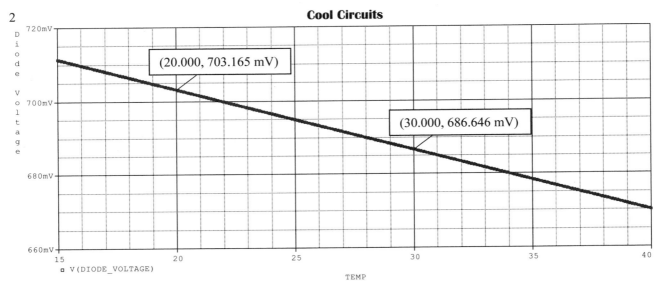

We see that the diode voltage appears to decrease linearly with temperature. Two points have been selected to obtain numerical values. These points show that at 20°C the diode voltage is 703 mV, and at 30°C the diode voltage is 686 mV. This corresponds to an average temperature coefficient of −1.7 mV/°C, or the diode voltage decreases by 1.7 mV for every 1°C increase in diode temperature. This is referred to as a negative temperature coefficient.

A current source is constructed of several components and can be a complicated circuit. Instead of using a current source, we will use a circuit that approximates a current source. In the circuit shown below, the current source has been replaced by resistor R_1. We will refer to this circuit as the resistor-based temperature sensing circuit.

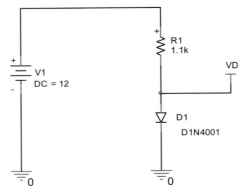

The current through the resistor and the diode is $I_D = \dfrac{V_1 - V_D}{R_1}$. Since $V_1 \gg V_D$, $I_D \approx \dfrac{V_1}{R_1}$ and the diode current is approximately independent of the diode voltage. Thus, as the diode voltage changes with temperature, the diode current remains approximately constant and approximately independent of temperature. For the current-source-based temperature sensor circuit, as the diode voltage changes, the current is constant and independent of temperature. We will compare the performance of these two circuits using PSpice.

The PSpice simulation below shows the diode voltage as a function of temperature for the two temperature sensing circuits. The lines appear nearly parallel indicating that both circuits have approximately the same temperature dependence. The reason that they are separated is that the current through each diode is slightly different because the 1.1 kΩ resistor allows 10.2 mA of current to flow through the diode while the current source is set to 10 mA.

Since the resistor closely approximates the performance of the current source, we would be foolish to use an expensive current source in this application when an inexpensive resistor will do the job.

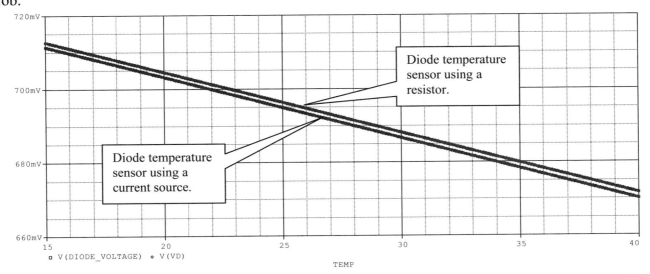

Now that we have the basic idea of how to use a PN diode as a temperature sensor, we will build our circuit. The premise of the circuit is that we will sense a temperature using a diode's voltage as an indication of temperature. We will compare this voltage to a fixed reference. When the temperature goes above a certain point indicated by the reference, a switch will turn on a fan that blows out the candle.

I.B. DC Power Supply

The first thing we need is a 12 V DC power supply to run our circuit. We will construct the regulated off-line supply below:

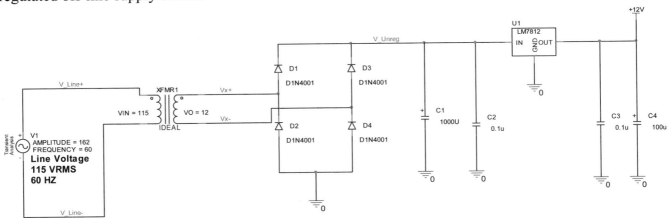

Note that we have shown large capacitors in parallel with small capacitors, such as $C_1 \| C_2$, and $C_3 \| C_4$. You will see this arrangement throughout the circuits shown in this manual. In the circuit above, C_1 and C_4 are large electrolytic capacitors that have a high series inductance. This inductance has high impedance at high frequencies, which means that electrolytic capacitors are not good capacitors at high frequencies. C_2 and C_3 are polyester or ceramic capacitors and have a low series inductance. We place them in parallel with the electrolytic capacitors so that the series inductance of the electrolytic capacitors is negated by the low impedance of the 0.1 µF capacitors at high frequencies.

We will now discuss the waveforms seen at intermediate portions of our DC supply. The voltage coming out of the wall socket is 115 volts RMS at 60 Hz. This is referred to as the line voltage, which is why we call the DC power supply shown above an off-line supply. 115 volts is an RMS average, and the actual voltage is a sinusoidal voltage with a peak value of $115 \times \sqrt{2}$, or

$v_{line}(t) = \left(115\sqrt{2}\right)\sin(2\pi60t) = 162\sin(2\pi60t)$. Thus, the line voltage is a sine wave with a peak voltage of approximately 162 volts. This is a much larger voltage than we would like to work with, so we will place this voltage through a 115 VRMS to 12 VRMS transformer. This transformer steps the voltage down to 12 volts RMS and also provides electrical isolation from the 115 V line. You can see from the circuit drawing that there is no direct wire connection between the circuit and the 115 V line. Power is passed from the 115 V line to the 12 V secondary by the magnetic field inside the transformer core. This type of isolation is called galvanic isolation. The transformer provides isolation and also allows us to work with a reasonably small voltage at the transformer's output (secondary winding).

The secondary of the transformer is a 12 V RMS sine wave at 60 Hz. Once again, the peak of this sine wave is the RMS value times $\sqrt{2}$, so the transformer output voltage is $v_{tran}(t) = \left(12\sqrt{2}\right)\sin(2\pi60t) = 16.97\sin(2\pi60t)$, or approximately a 17-volt sine wave.

The output of the transformer is connected to a full-wave rectifier (diodes D_1 through D_4). This arrangement of diodes is also referred to as a diode bridge rectifier. If this bridge were not hooked up to capacitors C_1 and C_2, the diode bridge would act like an absolute value circuit. To illustrate this part of the circuit, we will show waveforms of the circuit without the filter capacitors C_1 and C_2 and add a resistive load, R_1. The circuit below simulates the transformer and full-wave rectifier. The top trace shows the line voltage with a peak of about 162 volts. The middle trace is the transformer output with a peak of 17 volts, and the bottom trace is the output of the rectifier.

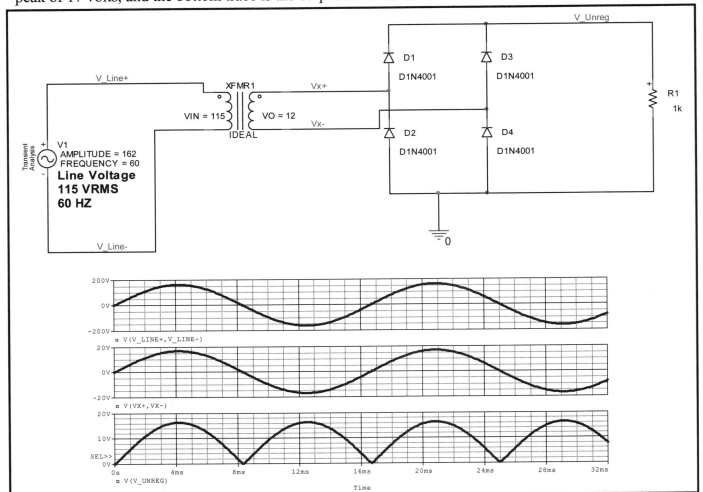

The output of the bridge rectifier can be considered a DC voltage with a large amount of ripple. To reduce the ripple, we add the filter capacitors and then the linear regulator (C_1, C_2, and U_1 in the complete DC supply on page 3). Without the filter capacitors, C_1 and C_2, the bridge rectifier would have too much ripple for the regulator to function properly. The purpose of the filter capacitors is to smooth out the ripple enough for the linear regulator to function. As long as the input to the linear regulator is greater than about 14 V, its output will be a very smooth DC voltage. We will now simulate the complete supply shown on page 3. For this simulation, we will draw 500 mA from the supply by connecting the output of the linear regulator to a 24 Ω load resistor. The waveforms are shown below:

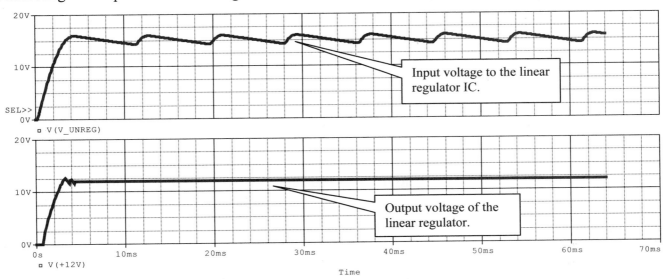

The top trace is the voltage output from the bridge rectifier with filter capacitors. We see that the ripple has been greatly reduced by the filter capacitors C_1 and C_2, but it has not been completely eliminated. The top waveform, however, is the input voltage to the linear regulator, and it is large enough for the regulator to function.

Here's the way the LM7812 linear regulator functions: If the input voltage is greater than approximately 14 volts and less than 35 volts, the output voltage will be held constant at 12 volts. Even if there is a lot of ripple at the input to the regulator, as in the top waveform above, the output is held constant as long as the input is greater than about 14 V. A specification for linear regulators called the ripple rejection ratio relates the ripple on the input that passes through to the output:

$$\text{Ripple Rejection (dB)} = 20\log_{10}\left(\frac{\Delta V_{IN}}{\Delta V_{OUT}}\right)$$

A typical number for the ripple rejection ratio is 70 dB. This means that a ripple of 5 volts at the input of the regulator would produce a ripple of less than 1.6 mV at the regulator output.

In the waveforms shown above, the input voltage to the regulator (top trace) has a large amount of ripple, but is always above 14 volts. Since the input is large enough, the regulator produces a smooth DC output voltage that we can use for the remainder of our circuit.

I.C. Temperature Sense Comparator

Now that we have a DC power supply to run our circuit, we can get down to the main business of building the components that perform the advertised functions of the circuit. The first component is the temperature sensor and comparator. We will use the resistor-based temperature sensor discussed in section I.A in the circuit below. Note that R_1 has been changed to 10 kΩ in the circuit below, while in the resistor-based temperature sensing circuit, we used $R_1 = 1.1$ kΩ. This change was made to reduce the static power dissipation of the resistor and diode. The consequence is that the diode voltage starts at a slightly lower value because the current is lower.

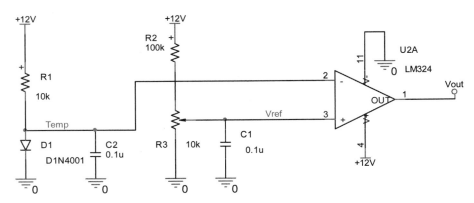

The voltage at node Temp is a function of the temperature and decreases as the temperature increases. At room temperature, the voltage is approximately 600 mV and it decreases by about 2 mV/°C. Capacitor C_2 is added as a low-pass filter to remove any noise that may appear at that point in the circuit due to external sources. We would like to compare the temperature-dependent signal at node Temp to a fixed reference (Vref); due to device tolerances, however, we do not always know exactly what that reference should be. If we build several thousand temperature sensors using diodes with the same part number, the forward voltage of each sensor will be slightly different because the parts may be from different manufacturers or from the same manufacturer but a different production run. Tolerance in processing will give the devices slightly different characteristics. Thus, we will generate the signal Vref using a variable resistor known as a potentiometer. In the circuit above, R_3 is a potentiometer, or "pot" for short. Some people also refer to it as a trimmer potentiometer or "trim pot." A model for a 10 kΩ potentiometer is shown in Figure I-1. For a pot, the total resistance between terminals A and C is fixed (in this example at 10 kΩ). With a pot, the resistances R_{3a} and R_{3b} are free to range between 0 and the maximum resistance as long as the sum of the two resistances is equal to the maximum resistance. For our 10 kΩ pot, the following must be true:

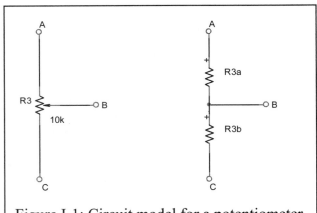

Figure I-1: Circuit model for a potentiometer.

$$0 \leq R3a \leq 10\,k\Omega$$

$$0 \leq R3b \leq 10\,k\Omega$$

$$R3a + R3b = 10\,k\Omega$$

We are using the pot as a voltage divider, which allows us to have a variable voltage set point for Vref.

Potentiometer settings are adjusted with a screw. There are 15-, 10-, and 1-turn pots, where the number of turns refers to the amount of the screw adjustment required to move the wiper terminal (B terminal in Figure I-1) from terminal A to terminal C. We need the voltage to vary by only a few millivolts. Had we hooked the pot directly to the +12V supply, Vref would be variable between 0 and 12 volts, and we would use only a fraction of this range (close to 0.6 volts). To make the pot adjustment more accurate, we have added R_2 in series with R_3 in the circuit above. The minimum voltage at Vref is zero volts when the pot wiper is at point C (R_{3b} set to zero and R_{3a} set to 10 kΩ). The maximum voltage at Vref occurs when R_{3b} is set to 10 kΩ and R_{3a} is set to 0 Ω (or the wiper is at terminal A):

$$V_{ref_{max}} = 12\,V\,\frac{10\,k\Omega}{10\,k\Omega + 100\,k\Omega} = 1.09\,V$$

We will adjust our potentiometer to about 575 mV for our simulation. For the actual circuit, the pot will have enough range for the needed temperature range, and we will set V_{REF} experimentally.

The operation of the comparator is fairly simple. At room temperature, the voltage at the Temp node is higher than the voltage at the Vref node. For the comparator, this means that the voltage at the negative (-) node is higher than the voltage at the positive (+) node, so the output of the comparator is low. As the diode heats up due to an external heat source, the voltage at the Temp node decreases. When the temperature becomes high enough, the voltage at node Temp will go below the voltage at node Vref, the comparator's + node will be higher than the - node, and the comparator's output will go high. Later on, we will use this high output to turn on a fan and blow out a candle. A simulation of the temperature sensor, reference voltage, and comparator output is shown below:

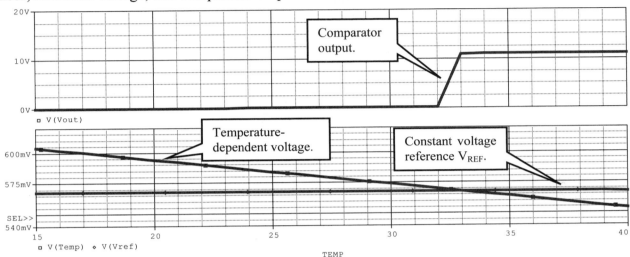

I.D. Transistor High-Current Driver

We would like to turn on and off a high-current fan with the output of our LM324 op-amp. The fan we would like to use requires 150 mA of current at 12 VDC, while a typical op-amp can only drive between a 5 mA and 20 mA load. To drive this large load, we will use a bipolar junction transistor (BJT) as a high-current driver. We will use the circuit shown below:

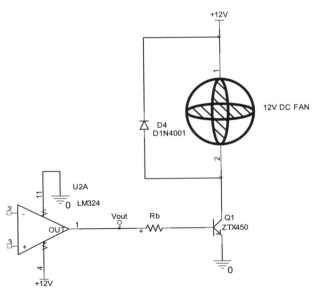

When a BJT is an ON switch, it is saturated and can be modeled as a voltage source of 0.2 or 0.3 volts. The fan will draw its rated current of 150 mA, which will be the collector current of our BJT. The maximum required base current we will ever need to drive this amount of collector current is $I_{B\max} = \dfrac{I_C}{h_{FE\min}}$. We chose a ZTX450 because at a collector current of 150 mA, the ZTX450 has a minimum value of h_{FE} of 100. This means a maximum base current of 1.5 mA is required to achieve a collector current of 150 mA for a ZTX450. A 2N3904 BJT has a minimum value of 30 for h_{FE} at $I_C = 100$ mA, and this requires too high a base current to be practical.

We must now choose resistor R_b. For an LM324, the maximum positive output cannot reach the positive supply of 12 volts. For design purposes, we will assume that the maximum positive output is 10 volts. This is a bit conservative, but if the base current is large enough when the output is 10 volts, the base current will be larger when the output is higher, which is OK because the BJT switch will be more deeply saturated. When the output is high, we have the situation shown in the picture below:

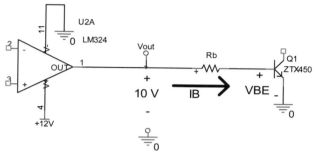

The base current is $I_B = \dfrac{V_{OUT} - V_{BE}}{R_b}$. R_b is the only unknown, so we can solve this equation for R_b:

$$R_b = \frac{V_{OUT} - V_{BE}}{I_B} = \frac{10\,V - 0.7\,V}{1.5\,mA} = 6200\,\Omega$$

We will round R_b down to the next smallest standard 5%, or 5.6 kΩ. All of the approximations we have made, including rounding down R_b, tend to make the base current I_B larger. This is OK because this causes the BJT to be more saturated when it is on. The more a transistor is saturated when it is an ON switch, the lower the saturation voltage, and the closer the BJT is to an ideal switch, which has an ON voltage of zero.

We have completed the design of our circuit; however, one last component requires explanation. Diode D_4 in our driver circuit is called a freewheeling diode, and is used for inductive loads such as a fan. [2] When the switch is on, approximately 12 V appears across the fan and the diode. The diode is reverse biased, so no current flows through it. Current flows through the fan, which we will model as a resistor and inductor in series in the figure below. We have replaced the BJT switch with a DC voltage V_{CEsat}, the model for a BJT when it is an ON switch.

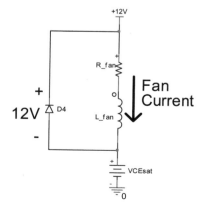

When the switch is ON, D₄ does nothing. The purpose for D₄ is when the switch turns OFF. If you remember, you cannot change an inductor's current instantaneously. If we did not have D₄, and we tried to turn off the BJT, we would be attempting to make the inductor current go from some non-zero DC value to zero instantaneously. The inductor would resist this change and develop a voltage large enough to break down the BJT. To avoid this problem, we provide a path for the inductor current to flow when the switch turns OFF, as shown below. No current flows through the switch because it is OFF, but current does flow through the freewheeling diode D₄.

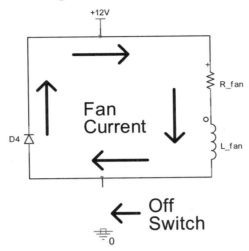

Current will flow through the freewheeling diode until the energy stored in the inductor is dissipated in the resistor and the diode. We see that the freewheeling diode is necessary only during the turn-off transient, and is inactive at all other times.

We have completed the discussion of this circuit. The complete circuit and a picture of the device are shown below and on the following page.

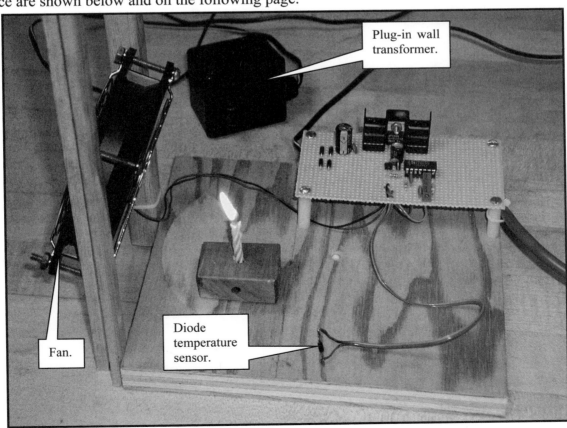

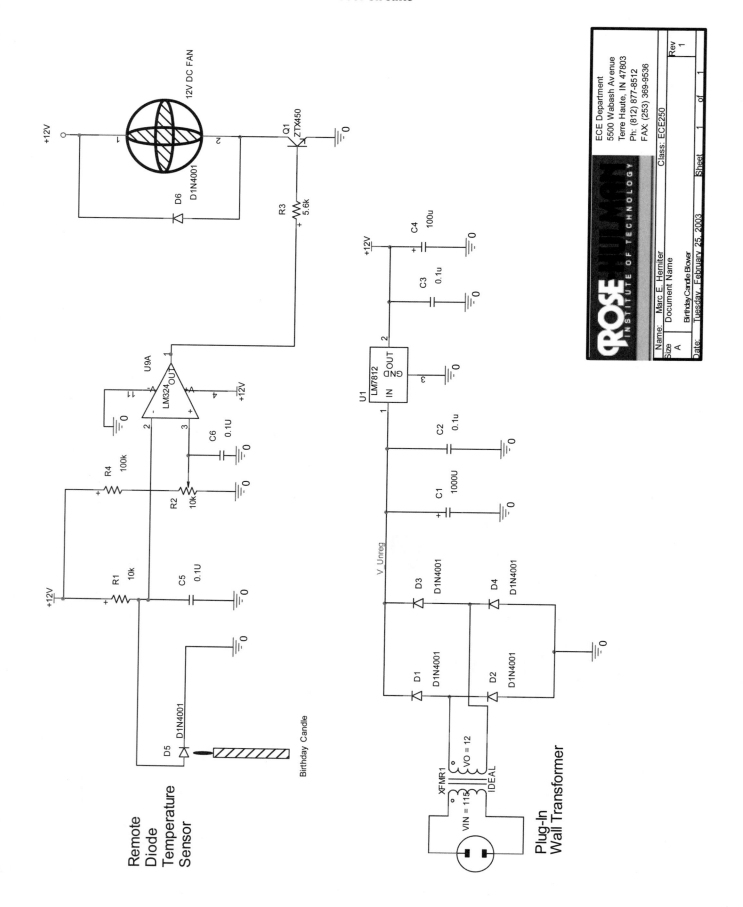

Cool Circuit II
Infrared Bug Sucker

I used to live in Flagstaff, Arizona. Arizona doesn't have too many bugs. There are insects, but not many insects that really "bug" you. Flagstaff is at an altitude of 7000 feet and is very dry, so there are few mosquitoes and flies. There are a few cool arachnids like tarantulas and black widow spiders, but they don't really bother you. One day I was stacking wood and I turned over a log and saw a huge black widow spider huddled in a groove. The round part of the spider must have been at least an inch in diameter. The spider must have been on steroids because most of the black widows I saw were quite small. We took one look at each other and ran in opposite directions. I'm not sure who ran faster; that spider took off pretty quickly. Black widow spiders are the ultimate fear for people like me who are afraid of spiders. And they are really black! You've never really seen black until you see a black widow spider. I wish I could have a car as black as a black widow spider.

Black widows are reclusive and hide quickly when they see you, so you don't notice them that much. Nonetheless, I was happy that I had not been bitten by the time I moved to Indiana. (Now I'm afraid of tornados and ticks. The wimpy spiders here don't bother me too much.) The first day we moved to Indiana, I opened the garage door, turned on the lights, and then walked into the house to take a look. When I returned to the garage, the formerly white ceiling was covered with so many bugs that it looked black. The entire ceiling was covered, not just near the lights, and the ceiling was slowly moving. That was my first introduction to bugs in Indiana. Since then I have learned three things about Indiana: if it is metal it rusts, if it is green it grows, and if it is a bug it will get in your house, fly around your face, and get in your food.

One day, while several flies were buzzing around my face and attempting to land in my salsa, I got completely annoyed, broke out the vacuum cleaner, and starting chasing them around the house in an attempt to catch them. Vacuum cleaners are the preferred way to catch black widow spiders because you can use a ten-foot hose and suck them up without getting anywhere near them. Once you get one, you drop the hose and leave the vacuum running for at least ten minutes just to make sure that the spider will not get out of the bag. (You drop the hose to prevent the rare quantum event of the spider tunneling through the hose and biting you.) The only down side to this method is that you have to occasionally empty the bag, and it is kind of creepy when you pull out the bag knowing that it is full of black widows, any of which could come back to life, crawl out of the bag, and then come after you. (After being dead, a black widow that comes back to life is much less reclusive.) It never happened to me, but I always use gloves to empty my vacuum cleaner bags. Anyway, chasing around flies with a vacuum cleaner does not work too well, and you don't want to let your neighbors see you doing it. Nonetheless, the seeds of an invention have been born. What if we could attract flies to a specific location, sense the presence of the flies, and then turn on a vacuum cleaner when the flies are at the optimal location for being sucked in by the vacuum cleaner? The infrared bug sucker has been born.

We will use an infrared (IR) transmitter/detector pair to detect the presence of the fly. The transmitter will emit a constant infrared signal to the detector. When a fly crosses the path between the emitter and detector, the signal will be blocked and the output of the transmitter will change. This change will be detected and used to turn on a vacuum cleaner.

II.A. Infrared Emitters and Detectors

We will use an infrared LED for our emitter and an infrared phototransistor for the detector. Make sure that the wavelength of the radiation emitted by the emitter is the same as the frequency that your sensor can detect. Not all IR emitters and detectors work at the same wavelength, so you need to check the datasheets to verify the wavelength at which the devices are designed to operate. Some typical wavelengths are 880 nm and 940 nm.

From a circuit point of view, an infrared LED works just like a visible light LED except that you cannot see the light. LED stands for light emitting diode, and it behaves like a diode when used in a circuit in that current can only flow in one direction. There are many different types of diodes: ones that can pass large amounts of current, ones that can block large reverse voltages, ones that turn on and off fast, and ones that turn on and off slow. LEDs are fairly slow, low-power devices. You are probably familiar with LEDs that are used as indicator lights.

When an LED is used in the forward direction, current will flow through it. The diode will have a forward voltage drop of around 1.5 V to 2.5 V with a typical value being about 1.8 V. [3] This is much higher than the standard diode voltage drop of about 0.7 V. The circuit below shows a typical forward-biased LED.

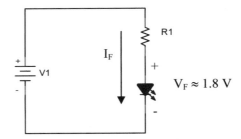

You will need to refer to the datasheet for your LED to find the range of values for its forward voltage.

Another major difference between LEDs and conventional diodes is that LEDs have a low reverse voltage rating. An applied reverse voltage is shown below:

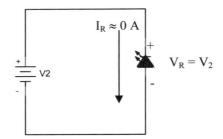

When voltage is applied in the reverse direction, the diode is supposed to be an open circuit, and no current should flow. No current will flow until the reverse voltage V_R becomes large enough to break down the diode, and then a large current can flow. The voltage at which current flows is called the reverse breakdown voltage. For rectifier diodes, typical breakdown voltages can range from 50 V to 1000 V depending on your needs. Rectifier diodes are designed to stop current from flowing in the reverse direction, and thus are designed to have a high reverse breakdown voltage. LEDs are designed to emit radiation and typically have low reverse breakdown voltage ratings. (A typical value is 5 V.) Thus, you can use LEDs to block only small reverse voltages. We will not use the LED to block voltage, but it is good to realize that it does have a low reverse breakdown voltage.

We will be using an infrared phototransistor to detect the infrared signal emitted by the diode. A phototransistor has only two leads: the collector and emitter. Instead of a base terminal, a lens focuses radiation on the base region. The incident radiation liberates charge carriers in the base region that act as the base current for the device: [4]

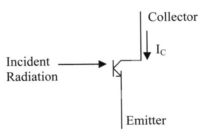

The collector current I_C is a function of the intensity of the incident radiation. Higher intensity yields more collector current.

For our IR diode and phototransistor, we will use the LIR204X and LPT2023. I do not know who manufactures them and they might not have the optimal characteristics for this application; but about 10,000 are available in our stock room, so we will use them because they are free and will accomplish the job. It is usually a good idea to use the parts that your company stocks rather than order new parts for every circuit you design. This does not mean that you should use inappropriate parts in a design. But if you can use a part that is already in stock or an assembly that was designed for another purpose in your design, you will save your company money. Plus, if the part is already in stock, your company probably knows a lot about how to use that part. Furthermore, your company probably has another engineer that has used the part and knows a lot about it. All of these are good reasons for using a stock part if it can be made to work in the design.

II.B. Bug Sensor

We will use the detection circuit shown below:

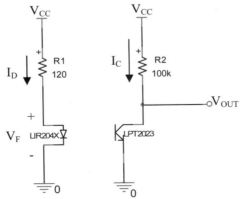

The IR diode (LIR204X) is biased so that it is always on and emitting IR radiation. The current through the diode is

$$I_D = \frac{V_{CC} - V_F}{R_1}$$

The maximum forward current rating of the diode is 50 mA, and the forward voltage V_F is specified to be between 1.2 and 1.6 V. Using this range in the forward voltage, $V_{CC} = 5$ V, and ignoring resistor tolerance, I_D will be between 31.7 mA and 28.3 mA. This value was chosen to emit a large amount of IR radiation, stay below the maximum rating of the device, and limit power dissipation in the LIR204X.

Mechanically, the IR diode points directly at the phototransistor so that the transistor is always illuminated. When the base of the phototransistor is illuminated, the incident radiation frees up charger carriers in the base region of the phototransistor and a collector current flows. The larger amount of incident radiation will yield a larger collector current. In the circuit above, the output voltage is $V_{OUT} =$

$V_{CC} - I_C R_C$. As the collector current increases, the output voltage decreases. Thus, for an increased amount of incident radiation, the collector current will increase and the output voltage will decrease.

When there is no obstruction between the IR diode and the phototransistor, V_{OUT} will be at its minimum value because most of the radiation is incident on the phototransistor. If nothing comes between the diode and phototransistor, V_{OUT} will remain constant at this low output voltage. But if a bug flies between the diode and the phototransistor, the bug will block some of the IR radiation and reduce the amount of radiation incident on the phototransistor. The bug will cause the collector current I_C to decrease, causing V_{OUT} to increase. Thus, if there is no bug between the diode and phototransistor, V_{OUT} will be at some low output voltage. If a bug crosses between the diode and phototransistor, the output voltage will increase. We have made a sensor that can detect the presence of a bug!

One issue that we have avoided is how to attract a bug to come between the diode and phototransistor. This problem, however, is a technical one that is beyond the scope of this text and is left to the reader as an exercise.

II.C. Comparator Circuit

Now that we have created a sensor whose output voltage varies with the presence of a bug, we need to create a circuit that monitors this voltage and makes a decision. The output voltage of the sensor may or may not vary significantly with the presence of a bug. It may vary by a few millivolts, or it may vary by several volts. We will determine this experimentally. Resistor R_2 was chosen empirically. The collector current is dependent on several factors, some of which are unknown. It depends on how much current flows through the emitter diode because the diode current determines how much radiation the diode emits. It depends on how far away the diode is from the phototransistor. It depends on how focused the emitted radiation is on the phototransistor. It depends on the conversion efficiency of the phototransistor, and many other factors. Thus, the resistor R_2 was made large enough so that V_{OUT} is fairly low (one or two volts) when no bug is between the diode and phototransistor. When a bug blocks the radiation, the collector current will decrease, and the output voltage will increase. The amount that it increases is dependent on how well the bug blocks the radiation. For a small change in I_C, however, there can be a large change in V_{OUT} if R_2 is large. Because $V_{OUT} = V_{CC} - I_C R_C$, the change in the output voltage due to a change in I_C is $\Delta V_{OUT} = V_{CC} - \Delta I_C R_C$. Thus, if we want a large change in the output of the sensor, we also need to choose R_2 to be large. R_2 was chosen experimentally by seeing how much the output voltage varied for a typical blockage between the diode and phototransistor. The upper limit on R_2 is such that if too large an R_2 is chosen, the phototransistor will become saturated. When a conventional BJT is saturated, to a first approximation a change in the base current does not change the collector current or collector-emitter voltage. For our phototransistor, even though there may be a small decrease in the incident radiation due to a blockage, the collector current and V_{OUT} will remain the same because the phototransistor is saturated. Thus, when the phototransistor is saturated, small changes in the amount of blockage will not change V_{OUT}. We want the output to change for any change in the incident radiation, so we do not want the phototransistor saturated.

Once we have chosen R_2, V_{OUT} will have a specific low value when no bug is between the diode and phototransistor. When a bug comes between the diode and phototransistor, the output voltage will increase above this low voltage. We will use a comparator to measure this voltage:

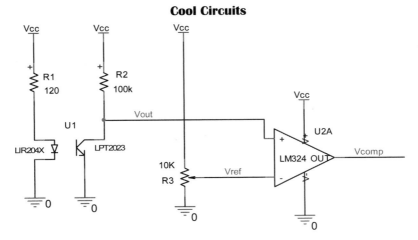

A comparator was discussed in Section I.C, and it is used in the same manner in this circuit as well. We will review its operation here. Resistor R_3 is a potentiometer (pot) and is used as a variable voltage divider. Potentiometers were discussed on page 6. Using the pot we can set the voltage Vref to any DC voltage we would like. The way the comparator works is if the voltage at the plus terminal (voltage V_{OUT} in this circuit) is greater than the voltage at the minus terminal (Vref in this circuit) then the output of the comparator V_{comp} goes high, which is as close as the comparator output can come to the supply voltage V_{CC}. Thus, if V_{OUT} > Vref, then $V_{comp} \approx$ high. On the other hand, if the voltage on the minus terminal is greater than the voltage on the plus terminal, then the output will go low, which is as close as the comparator output can come to ground. In summary:

$$\text{If } V_{OUT} > \text{Vref, then } V_{comp} \approx \text{high}$$

$$\text{If } V_{OUT} < \text{Vref, then } V_{comp} \approx \text{low}$$

When no bug is between the diode and phototransistor, V_{OUT} will be at its lowest value, which we will call V_{OUTL}. We want the comparator output to be low when there is no bug, so we will choose voltage Vref > V_{OUTL}. The desired sensitivity of the comparator will determine how close we want Vref and V_{OUTL} to be. Thus, when no bug is present, Vref > V_{OUT}, and the comparator output will be low. When a bug comes between the diode emitter and phototransistor, the output voltage will go up. The amount of blockage will determine how much V_{OUT} increases. If we want the comparator to flip for a very small amount of blockage, we will choose Vref close to V_{OUTL}. If we want the comparator to flip only when there is a significant amount of blockage, then we need to choose Vref further away from V_{OUTL}. The difference between V_{OUTL} and Vref is also determined experimentally. If you choose Vref too close to V_{OUTL}, then a small amount of noise on either comparator input could cause the comparator to flip. Conversely, if you choose Vref very far away from V_{OUTL}, then a typical bug may not block enough IR radiation to cause V_{OUT} to go above Vref, resulting in the comparator output never changing. At a minimum, with no bug present between the diode and phototransistor, Vref must be greater than V_{OUT}. When a bug goes between the diode and phototransistor, V_{OUT} must increase above Vref.

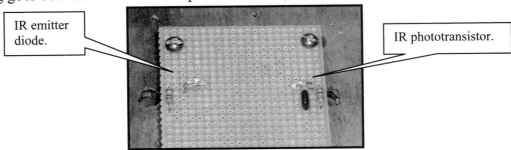

IR emitter diode.

IR phototransistor.

I adjusted Vref experimentally. I rolled up some little bits of paper that were about the size of a fly. (They kind of looked like the spitballs I used to shoot at my teacher in the eighth grade.) I would measure the comparator output with a meter or oscilloscope. I then placed a single paper ball between the diode and phototransistor and observed the output of the comparator. When the output flipped high or low due to the presence or absence of the paper ball, I decided that the value of Vref was appropriate. To adjust Vref, you will have to measure both V_{OUT} and Vref with and without the paper ball between the diode and phototransistor so that you know how V_{OUT} changes relative to Vref. You will not be able to just measure the comparator output and blindly change Vref in hopes of finding a suitable value. You will need to see how V_{OUT} varies with and without the blockage, and then adjust Vref appropriately.

II.D. Vacuum Cleaner Drive Circuit

When a bug is present, we would like to turn on a vacuum cleaner and suck up the bug. We will locate the hose directly over the bug sensor. When the comparator indicates that a bug is present, we will turn on a vacuum cleaner that will suck up the bug and remove the blockage between the IR diode and phototransistor. When the bug has been removed the comparator will flip to the opposite state. A picture of the vacuum hose and mechanical setup is shown below:

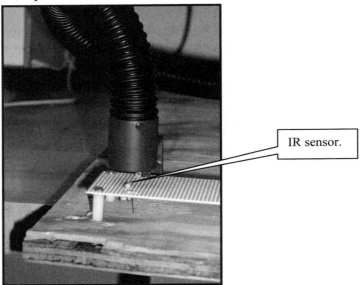

IR sensor.

A typical household vacuum cleaner runs off the 115 VAC line. We can control 115 VAC using MOSFETs, but the easiest way is to use a relay. A relay is an electromechanical switch that is controlled by passing a small current through the relay coil. A schematic drawing of a relay is shown below:

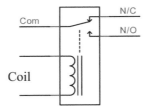

In the picture above, the relay contains a single switch that is controlled by the magnetic field of the coil. When no current is passed through the coil, there is no magnetic field and the common terminal will connect to the terminal labeled N/C, which stands for "normally closed." Thus, with no current through the coil, the connection between the Com terminal and the N/C terminal is a closed switch, or a short circuit. When we apply a current through the coil, a magnetic field is created that moves the arm of the switch to the N/O terminal. N/O stands for normally open, and the switch between the Com and N/O terminals is normally open when no current flows through the coil. When current flows through the coil, the Com terminal connects to the N/O terminal. We can use the relay to turn on something by connecting it to the Com and N/O terminals, turn off something by connecting it to the Com and N/C terminals, or switch the Com terminal between two different circuits.

We would like to turn on a vacuum cleaner that is powered by the 115 VAC line. The vacuum cleaner should normally be off, so when the relay is not energized (when no current flows through the coil) the vacuum cleaner should not be connected to the 115 VAC line. When we energize the coil, we want to connect the vacuum cleaner directly to the 115 VAC line. We will use the circuit below:

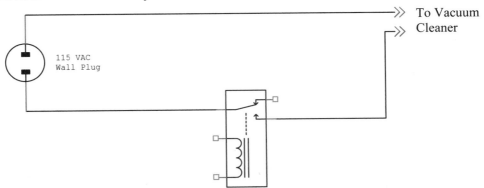

With this circuit we can turn the power to the vacuum cleaner on and off. When choosing a relay, we must make sure that the switch contacts (Com, N/C, and N/O in our example) are rated for the load we wish to control. In our example, the vacuum cleaner is the load and the contacts must be rated to deliver the power needed by the vacuum cleaner. There are high-voltage relays, relays for turning on and off 115 VAC, and low-voltage relays for turning off voltages of about 10 to 20 volts. Relays are also rated as to whether they can turn off AC or DC voltages. It is easier to turn off an AC voltage than a DC voltage because the current in an AC waveform goes through zero. The current in a DC waveform, on the other hand, continues to flow and you must interrupt that current by opening the switch contacts. You will note that if a relay has both DC and AC contact ratings, the contacts will be rated lower for DC voltages than for AC voltages. The contacts are also rated for their current carrying capability. When you choose a relay, it must be rated for both the current and voltage requirements of the load. The vacuum cleaner I am using is rated at 115 VAC and 10 A.

Now that we have chosen a relay whose contacts are rated for our load, we must look at energizing the relay. The relay coil will also have specifications. Some coils are energized with an AC voltage and some are energized with a DC voltage. If the coil is an AC coil, all you do is apply the required voltage of the specified amplitude and frequency. The coil will draw an amount of current specified in the datasheet. Some relay coils operate with 115 VAC and some operate at lower AC voltages like 12 VRMS or 24 VRMS. This voltage is usually available from a transformer that you have used in your circuit elsewhere.

We will be using a relay whose coil is energized with a DC voltage. To energize the coil, we apply the specified voltage. For example, we will use a relay with a 12 V coil in the circuit below:

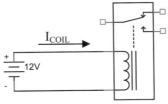

The coil will draw the amount of current specified in the datasheet. The coil is an electromagnetic device and is drawn as an inductor in the schematic, so it should behave like an inductor. That is, an inductor is a short to DC, so the coil current I_{COIL} should become infinite. So why does the coil not short out the DC supply? The question is a good one. We have not mentioned that the coil has a very large series resistance, and this resistance is what limits the current. A model for the coil is shown below, including the applied 12 V supply that energizes the coil.

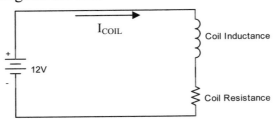

When we apply the 12 V, after the initial transient dies away, the current through the coil will be determined by the voltage and the coil resistance. Manufacturers will specify either the coil current or the coil resistance. If you know the specified voltage and the coil resistance, you can determine how much current the coil requires. Basically, we design the drive circuit for the relay coil as if it were a resistance.

There are two important things to remember about the coil of a relay. One is that the coil resistance determines the DC current through the coil. The second is that the coil does have inductance and the current through this inductance will not go to zero instantaneously. We will deal with the inductance by adding a freewheeling diode, as we did for the fan in the candle blower example.

The relay I will use operates at about 185 mA with a 5 VDC source. We will turn the voltage to the relay coil on and off using a bipolar junction transistor the same way we turned the fan on and off in section I. The design procedure for turning the fan on and off was covered on pages 7–8. The fan operated at 12 V and 150 mA. The design for the relay is exactly the same except that it uses 5 V and 185 mA, so we will not repeat the procedure here. The circuit for driving the relay is shown below:

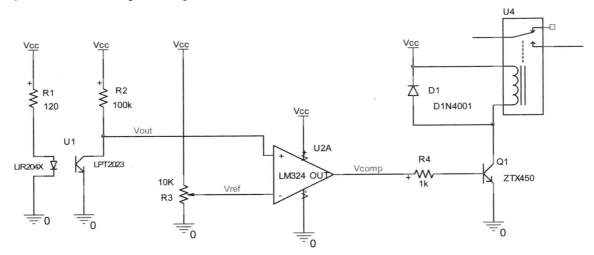

V_{CC} in the circuit is 5 V. Diode D_1 is a freewheeling diode that is used to clamp the inductive voltage spike that would occur across the coil inductance if we were to try to make the coil current go to zero instantaneously. The freewheeling diode was previously used in the candle blower to clamp the inductive voltage spike across the fan. See page 8 for a detailed discussion of a freewheeling diode.

II.E. Power Supply

We would like to run this circuit from the 115 V line, but it requires a 5 V DC power supply. We will create an off-line DC power supply using the method covered in section I.B. The complete power supply is shown below:

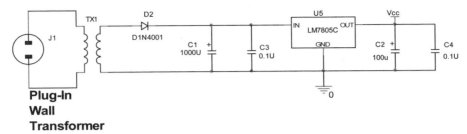

The circuit uses a plug-in wall transformer to convert the 115 VAC to 12 VAC. We use a half-wave rectifier to convert the 12 VAC to a DC voltage with a large amount of ripple. The LM7805C is a linear regulator that takes a DC input with a large amount of ripple and produces a 5 V output with very little ripple. A 12 V version of this supply was discussed in detail in section I.B, so we will not discuss it here.

II.F. Complete Circuit

A picture of the Bug Sucker and the complete circuit are shown below and on the following page.

Cool Circuits

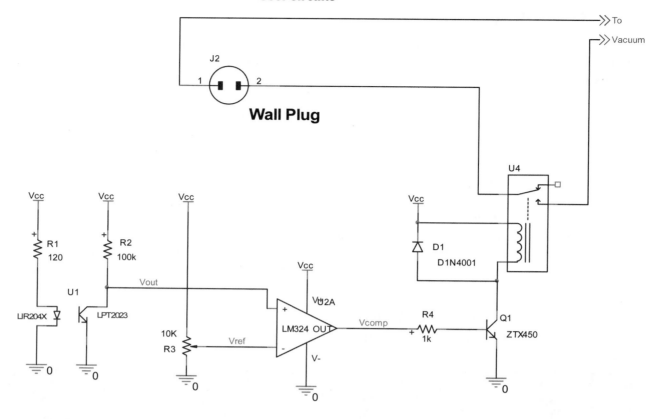

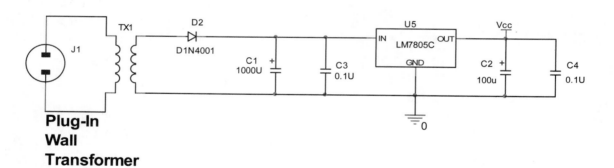

Cool Circuit III
Klingon Pain Stick

With this circuit we create a high-voltage DC source from a low-voltage sinusoidal source using a circuit called a voltage multiplier. In the area of particle physics, this circuit might also be called a Cockcroft-Walton accelerator and can be used to generate very high voltages. [5] Here, we will generate a voltage of about 300 volts from a 24 Vrms transformer.

The complete circuit is:

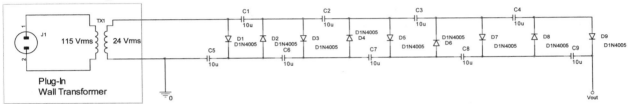

The output of the transformer is a 24 Vrms sine wave at 60 Hz. For this example, we can think of the transformer output as an ideal sinusoidal voltage source with a peak amplitude of about 34 volts. It turns out that the analysis of this circuit is quite complicated, so we will simplify our analysis and assume that the input waveform is a square wave with the same amplitude as the transformer output.

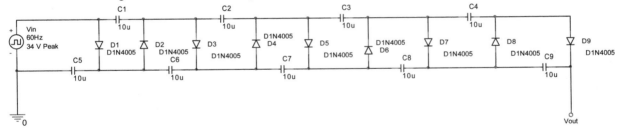

We will start by assuming that all capacitors are discharged and the diodes are ideal. When Vin is at its positive maximum, diode D_1 will be forward biased and all other diodes will be off. We will have the circuit below:

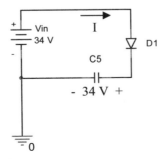

Current will flow in the direction shown and the capacitor will charge up to 34 V. When the input flips to –34 V, all of the diodes will be off except for D_2. For the moment, assume that $C_5 >> C_1$, so that C_5 behaves like 34 VDC source. We will have the circuit below:

21

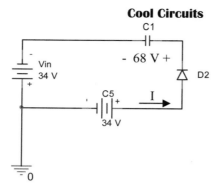

If C_5 were really much, much greater than C_1, then C_1 would charge up to 68 volts as shown. In our circuit C_1 equals C_5, so C_5 does not really behave as a voltage source. What does happen is that current flows in the direction shown, and charge is transferred from C_5 to C_1. Since charge is leaving C_5, the voltage on C_5 will go down. Since charge is flowing into C_1, its voltage will increase. Thus, after one cycle, the voltage on C_1 is less than 68 V and the voltage on C_5 is less than 34 V. During the next positive portion of the cycle, the voltage on C_5 is restored to 34 V. During the negative portion of the cycle, charge is again transferred from C_5 to C_1. The voltage on C_1 increases from the previous cycle and once again, the voltage on C_5 decreases below 34 V; however, it does not decrease as much as in the previous cycle because the voltage difference around the loop before the charge transfer is smaller than in the previous cycle and the charge transferred from C_5 to C_1 is less than in the previous cycle.

We see that during every positive portion of the cycle, the voltage on C_5 is restored to 34 V. During every negative portion of the cycle, charge is transferred from C_5 to C_1. Since there is no discharge path, eventually C_1 charges up to 68 volts after several cycles.

Next, let's assume that C_5 and C_1 are large compared to C_6 and behave like voltage sources. When Vin is positive, D_3 will be on and we will have the simplified circuit below:

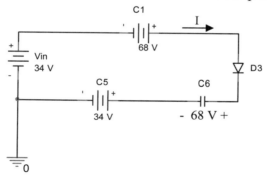

For C_1 and $C_5 \gg C_6$, C_6 will charge up to 68 V as shown. This is not really the case because all of the capacitors are equal. What really happens is that current flows in the direction shown and charge is transferred from C_1 to C_6. Thus the voltage on C_1 goes down and the voltage on C_6 goes up. During the positive portion of the cycle, the voltage on C_5 is also restored to 34 V through D_1 (not shown in the circuit above). During the negative portion of the cycle, charge is transferred from C_5 to C_1, which restores the voltage of C_1. For this three-capacitor circuit, charge is transferred from Vin to C_5. Then charge is transferred from C_5 to C_1. Finally, charge is transferred from C_1 to C_6. It takes several cycles for C_1 and C_6 to reach 68 V because the charge that was originally transferred from Vin is shared between C_1, C_5, and C_6.

Next, let's assume that C_5, C_6, and C_1 are large compared to C_2, and behave like voltage sources. When Vin is negative, D_4 will be on and we will have the simplified circuit shown below:

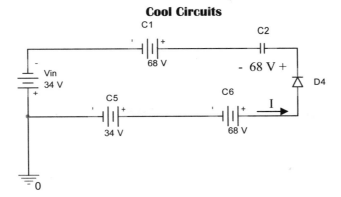

Using the same reasoning as before, after several cycles, the voltage on C_2 will approach 68 V. Charge is transferred from C_5 and C_6 to C_1 and C_2 around multiple loops. One loop is the loop shown above. There is, however, another loop that transfers charge from C_5 to C_1 through D_2, not shown above.

The analysis is much more difficult than we have discussed here because of the charge sharing, because the source is actually a sine wave, and because of the multiple charging loops. We conclude, however, that if the input is a sinusoidal source with amplitude Vx, then C_5 charges up to voltage Vx and all of the other capacitors charge up to 2·Vx:

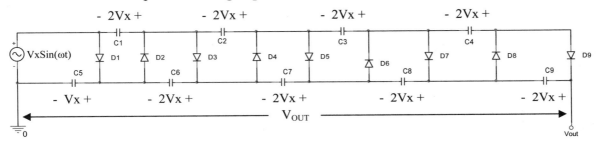

The output is taken between the V_{OUT} terminal and ground, so for the circuit shown, $V_{OUT} = 9Vx$. For a 24 Vrms transformer, Vx = 34 volts, so the output of this circuit is about 306 VDC. Note that we can keep repeating the structure, so output voltages much higher than 306 V are possible.

It takes several cycles for charge to reach C_9, and due to charge sharing, it takes several cycles for each capacitor to reach 2 Vx. To see how long it actually takes to charge up this circuit, we will run a PSpice simulation. The output voltage is shown below:

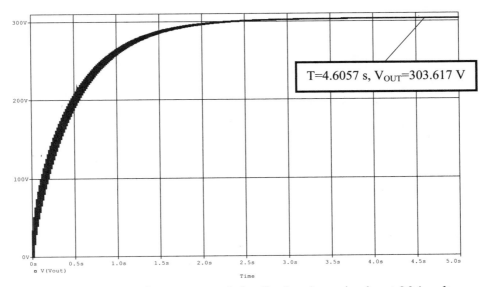

The output takes about 3 seconds to charge up, and the final voltage is about 304 volts.

Some notes about this circuit are in order. Because the circuit takes such a long time to charge up, you should not expect to draw much continuous power from the circuit. Attaching a 300 kΩ load to the output reduces the DC output voltage of the circuit to about 230 volts. Because the circuit is so sensitive to load current, low-leakage capacitors must be used. If you remember, capacitor leakage can be modeled as a resistance in parallel with the capacitor. [6] If we used electrolytic capacitors, which have a high leakage current, to construct this circuit, the parallel resistance of the capacitors would behave just like a load resistance and tend to reduce the output voltage. So for best performance, you should use low-leakage capacitors like ceramic or polyester.

So why did I call this circuit a Klingon™ pain stick? The topology of the circuit suggests that the components be laid out in a long narrow pattern. When I built it, I mounted it on a stick with the two electrodes at the end. You discharge the circuit by touching it to a conductor. If the material you touch has a low enough resistance, a large spark will occur. The shape and the energy discharge (spark) reminded me of the Klingon pain sticks shown in Star Trek: The Next Generation™. Pictures of my realization are shown below. This circuit has 19 capacitors and an output of about 600 V.

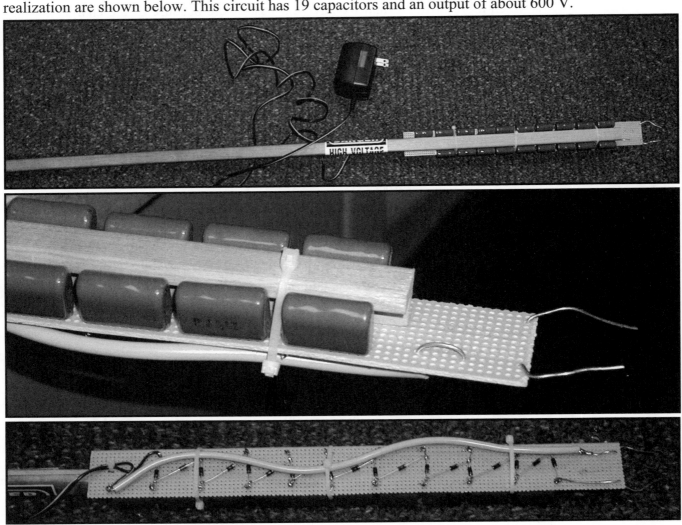

Cool Circuit IV
Stun Gun (18 V to 1000 V Boost Converter)

This circuit is a boost switching regulator that steps up two 9 V batteries in series to about 1000 V DC. The basic circuit is shown below:

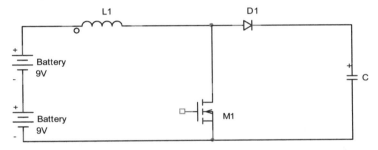

M1 is a MOSFET that we will use as a switch. When the switch is closed, 18 V is placed across the inductor. The diode is reverse biased because the capacitor voltage will be positive for this topology:

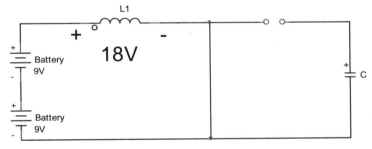

Remember that the governing equation for an inductor is $V_L = L\dfrac{dI_L}{dt}$. We can solve this equation for the inductor current:

$$I_L(t) = \frac{1}{L}\int_0^t V_L(t)dt + I_L(0)$$

For our circuit, $V_L(t)$ is constant at 18 V and we will assume that the initial condition on the inductor is zero, $I_L(0) = 0$. In the world of switching power supply design, this means that this boost converter is operating in the discontinuous mode of operation. [7] Using these conditions, our equation becomes:

$$I_L(t) \quad = \quad \frac{1}{L}\int_0^t V_L(t)dt + I_L(0)$$

$$= \quad \frac{1}{L}\int_0^t (18\,\text{volts})dt$$

$$= \quad \left(\frac{18\,\text{volts}}{L}\right)t$$

This is an equation of a straight line with slope $\left(\dfrac{18\,\text{volts}}{L}\right)$. Thus, when the switch is closed, the current in the inductor will increase as a straight line. An example with a 1 mH inductor is shown below:

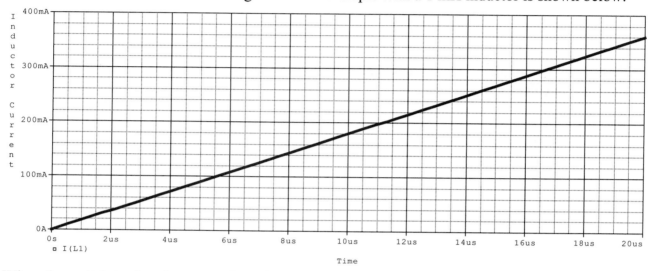

When the switch is closed, current flows in the loop as shown below:

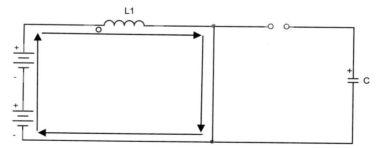

At any point in time, the amount of energy stored in the inductor is $E_L = \frac{1}{2}LI^2$. When we open the switch, the current in the inductor cannot change instantaneously, so it continues to flow through the diode and capacitor:

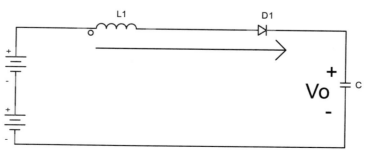

This current charges the capacitor voltage Vo in the polarity shown. The energy that was stored in the inductor's magnetic field ($E_L = \frac{1}{2}LI^2$) is transferred to the capacitor and stored in the capacitor's electric field ($E_C = \frac{1}{2}CV^2$). To store this extra energy, the capacitor voltage must increase.

Every time we close the switch, we take energy from the battery and store it in the inductor in the form of magnetic energy. When we open the switch, the energy is transferred from the inductor to the capacitor. As we repeat this process, we move energy from the battery to the capacitor. In order to store more energy in the capacitor, the capacitor's voltage must increase.

One point that we didn't mention is that when the switch is open and the inductor discharges into the capacitor, the voltage across the inductor reverses and the inductor current decreases. To see this more clearly, we will look at an example. Suppose the capacitor voltage is 50 volts and the switch is closed as shown below:

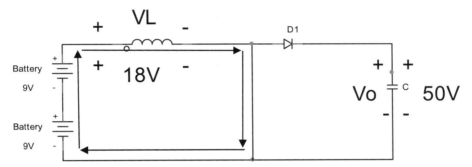

The diode is off because it is reverse biased with 50 volts. The inductor has a constant voltage of $V_L = $ 18 V across it, and the current will increase linearly because the current is $I_L(t) = \dfrac{1}{L}\displaystyle\int_0^t (18 \text{ volts})dt$. When the switch opens, current must flow through the diode and into the capacitor. We will assume that the diode is ideal so that when it is on, we replace it by a short circuit. Thus, when the switch is off and current flows through the diode, we have the circuit below:

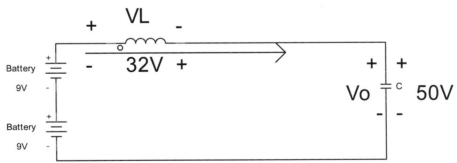

Because the diode is on, the inductor is now connected between the capacitor and the batteries. Adding up the voltages around the loop we see that $V_L = -32$ V. The equation for calculating the inductor current is the same as before except that the voltage is negative, so the ramp will be decreasing rather than increasing. Thus, when we close the switch, the inductor starts at zero and increases linearly with a slope of $\left(\dfrac{18 \text{ volts}}{L}\right)$. When the switch opens, the voltage across the inductor flips to –32 volts and the current decreases linearly with a slope of $\left(\dfrac{-32 \text{ volts}}{L}\right)$. In general, when the switch opens, the voltage across the inductor will flip to $(18\,\text{V} - V_O)$ and the current will decrease with a slope of $\left(\dfrac{18\,\text{V} - V_O}{L}\right)$.

The waveforms shown below illustrate the process. First, we will show the inductor current and voltage for the circuit below. The capacitor voltage is fixed at approximately 50 volts:

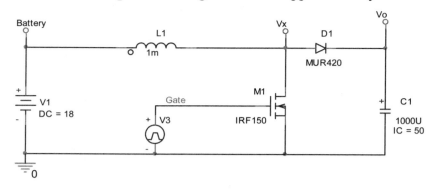

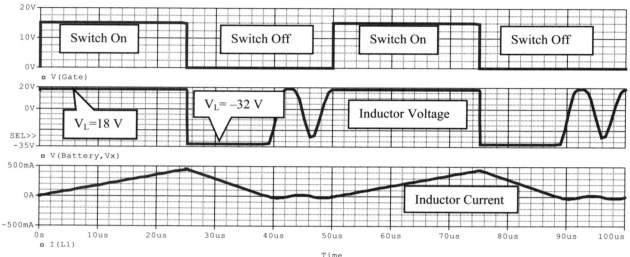

We see that when the inductor voltage is positive, the inductor current ramps up. When the inductor voltage is negative, the inductor current ramps down. We also see that when the inductor current goes to zero, the inductor current flip-flops around. This is because both the diode and switch are off and the inductor voltage is free to vary wildly.

Next, we will start the capacitor voltage at 50 volts, let the simulation run for several cycles, and watch the capacitor voltage increase. We will use the same circuit with a smaller capacitor so that the voltage can grow more rapidly. A large capacitor will also work, but it will take more charge-discharge cycles to build up to the same voltage. To see the voltage build up in a few simulation cycles, we will use a 100 nF capacitor.

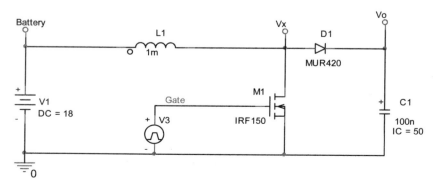

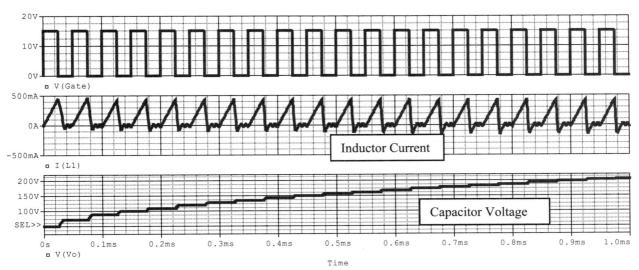

We see that the capacitor voltage increases each time energy is dumped into it. To see the steps more clearly, we will zoom in on the traces a bit:

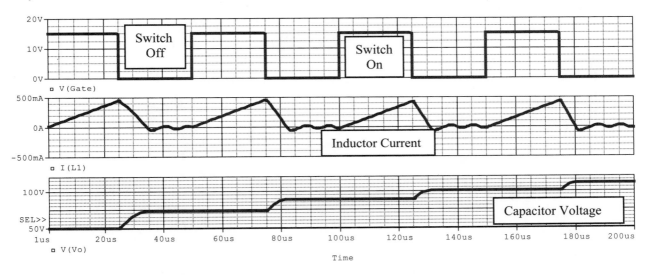

With an ideal circuit, we could repeat this process indefinitely and obtain infinitely high voltages. In reality, all components have voltage and current limitations. In practice, the maximum output voltage is limited by the maximum voltage limits of the components we use to construct our circuit. The maximum output voltage of our simulation is limited by the MUR420 diode, which is a 200-volt diode. If we let the simulation run much longer, the diode will have a reverse voltage greater than 200 volts. At this point the diode will break down and no longer be an open switch.

IV.A. Complete Circuit

The complete circuit is shown on page 31. We will briefly describe a few of the extra elements that were not mentioned earlier.

- R_3 and D_2 form an ON-OFF light. The LED illuminates when the converter is on. (An important feature for a circuit that produces 1000 V!)

- D_1 is a UF1007 and has a reverse breakdown voltage of 1000 volts. This diode is similar to a 1N1007 except that it is an "ultrafast" diode. The 1N1007 diode is a "rectifier" diode, which means that it is designed to work at low frequencies, typically 60 Hz. Our boost converter has a

20 kHz switching frequency and requires D_1 to turn off in a few microseconds. Thus, an ultrafast diode is needed.

- The output voltage is designed to reach 1000 volts. The largest 1 µF capacitor we could find was a 630 V capacitor. If you place two 1 µF capacitors in series, the DC voltage across the capacitors will not split evenly across the capacitors because each capacitor has an internal leakage resistance that can be modeled as a parallel resistance. The leakage resistance of the two capacitors is not likely to be the same, and the voltage across the two capacitors will split according to the leakage resistance and not the capacitance values. To evenly split the voltage across the two capacitors, we have added external voltage grading resistors, R_1, R_2, R_4, R_5, and R_8, to evenly split the voltage across C_1 and C_4.

- The UC3525A is a pulse-width-modulating (PWM) control chip, and it controls the MOSFET switches. Voltage Vsig is the voltage across resistor R_8. Using a voltage divider relationship, Vsig is approximately the output voltage divided by 350. Vsig is fed back to the UC3525A control chip. When the output voltage is high enough, switching is halted and capacitors C_1 and C_4 hold their voltage. If the output voltage decays below a specified point, switching starts again, and the output voltage is boosted back up to a higher value.

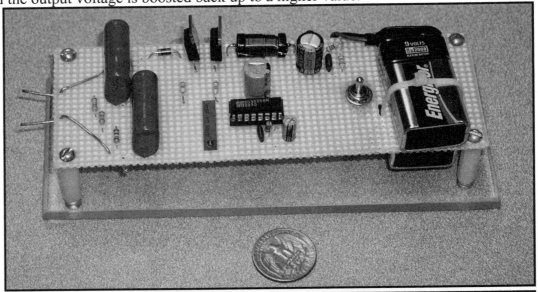

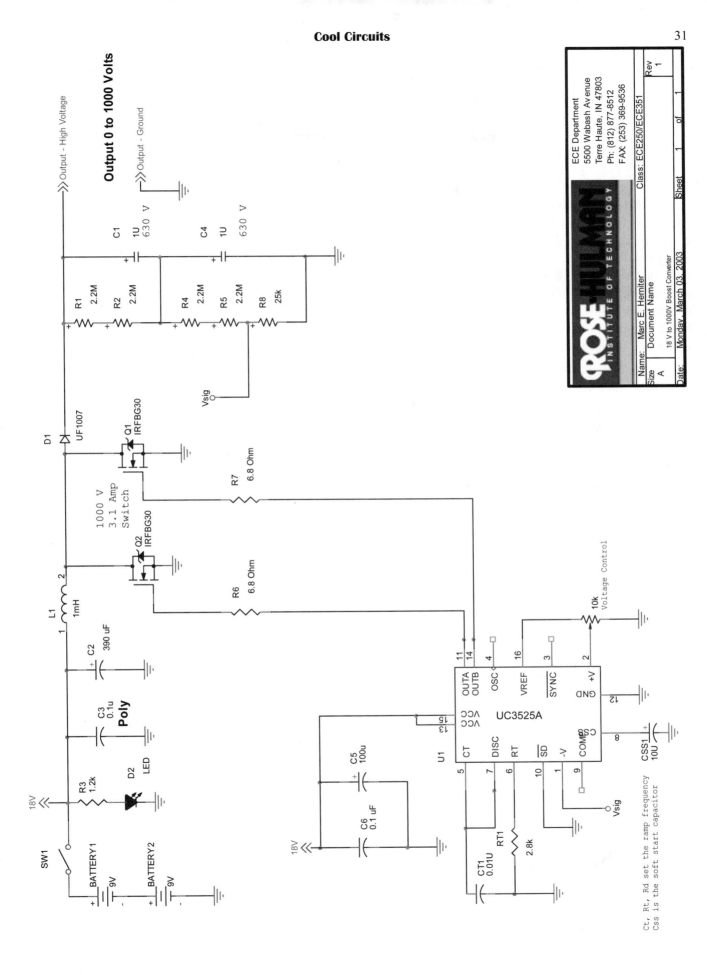

Output 0 to 1000 Volts

Output - High Voltage

Output - Ground

ECE Department
5500 Wabash Avenue
Terre Haute, IN 47803
Ph: (812) 877-8512
FAX: (253) 369-9536

Name: Marc E. Herniter
Class: ECE250/ECE351
Document Name
18 V to 1000V Boost Converter
Size A
Date: Monday, March 03, 2003
Sheet 1 of 1
Rev 1

Ct, Rt, Rd set the ramp frequency
Css is the soft start capacitor

Cool Circuit V
DC Motor Speed Controller

An area in electronics that is becoming more and more important is the field of power electronics where we can control a large amount of power with low-voltage, low-power-control signals. Power electronics includes DC to DC converters, DC to AC inverters, motor controllers, battery chargers, and many other devices. A DC to DC converter converts a DC or constant voltage to another DC voltage. An example is the 18 V to 1000 V boost converter discussed in Section IV. DC to AC inverters convert a constant voltage into an AC waveform, typically a square wave or a sine wave. An example is an uninterruptible power supply (UPS). Some UPSs will convert 12 VDC to a 115 VAC RMS square wave. More expensive UPSs will convert small DC voltages provided by batteries into 115 VAC RMS sine wave.

In this section we will look at using power electronics to control a motor. The techniques discussed here could be used to control a motor for a low-power toy RC car, or for an electric vehicle that has a peak power over 100 kW.

Suppose that we have a DC motor and we want to control its speed. The motor is designed to operate at its name plate voltage of 200 V. The current that the motor draws depends on the mechanical load attached to the motor. The speed of the motor depends on the mechanical load and the applied DC voltage. We will assume that the mechanical load is constant for this example. To control the speed, the only thing we need to do is change the DC voltage applied to the motor. Thus, we can control the speed by making a variable DC voltage source:

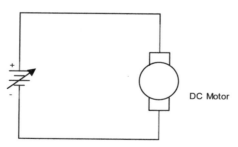

DC Motor

For low-power, low-voltage motors, this method may be possible. It is fairly easy to build a variable DC source using a voltage regulator such as the LM317. Note that linear voltage regulators are very inefficient and a large fraction of the battery power will be wasted in heating up the linear regulator. Suppose that our motor ran on a peak voltage of 200 V and we wanted to supply a maximum current of 600 A? Building a variable DC supply of this magnitude is non-trivial. (A huge understatement!) It is fairly easy to build a fixed DC supply that can supply 200 V at a maximum current of 600 A using lead acid batteries. So, how do we take a fixed voltage and make it appear to be a variable voltage for the motor?

V.A. Pulse-Width Modulation

We will use pulse-width modulation (PWM) and a switch to make the average voltage across the motor vary between 0 and the maximum DC voltage. Suppose we have a switch in series with the motor that we can open and close, such as the circuit shown below:

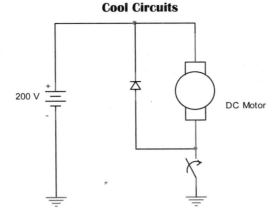

When the switch is closed, the voltage source is connected directly across the motor, and the motor voltage is equal to the source voltage of 200 V. The voltage across the diode in the reverse direction is also 200 V, so the diode is off and no current flows through it. Thus, when the switch is closed, we have the circuit below:

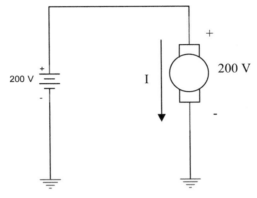

A motor has a large amount of inductance. When the switch opens, the motor current cannot go to zero instantaneously, so the diode is added to provide a path for current to flow:

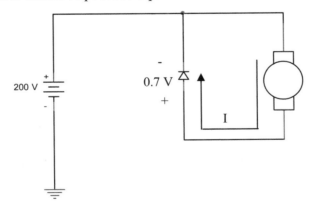

The diode will be forward biased and will have a small forward voltage of about 0.7 V. This diode is referred to as a freewheeling diode. This voltage has the opposite polarity of the DC voltage source and is approximately zero. Thus, when the switch is on, the motor voltage is 200 V, or equal to that of the DC source. When the switch is off, the motor voltage is approximately zero.

If we open and close the switch at a regular frequency, the voltage across the motor will be a 200 V square wave:

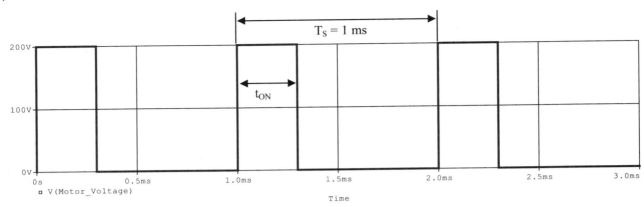

The waveform repeats at a regular interval called the switching frequency F_S, which is the frequency at which the switch is opened and closed. The switching frequency is the reciprocal of the switching period T_S, which is shown on the waveform as 1 ms. Thus, the switching frequency for this waveform is $T_S = 1/F_S = 1$ kHz. When the switch is closed, the voltage across the motor is 200 V. The amount of time the switch is closed is labeled as t_{on} in the waveform. We will be using a transistor as a switch for this circuit. When the transistor is "on," it is modeled as a closed switch, which gives rise to the label t_{on}.

We will use pulse-width modulation (PWM) to control the motor. With PWM control, we will keep the switching frequency constant and vary the on time. An example is the waveform below:

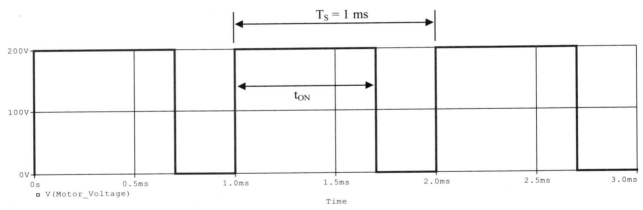

We see that the switching period T_S is the same as in the previous waveform, but the on time has increased. The on time (t_{on}) can range from 0 to 1 ms. When $t_{on} = 0$ s, the switch is off all the time and the motor is off. When $t_{on} = 1$ ms, the switch is on all the time, and the motor is at maximum power.

The waveforms shown will be the voltage across the motor, which is a mechanical device with a large mass. The rotor of the motor cannot change its speed at a frequency of 1 kHz and responds to the time-average of the waveform. We can calculate the time-average of the waveform as: [8]

$$\left\langle V_{Motor\_Voltage} \right\rangle = \frac{1}{T_S} \int_0^{T_S} V_{Motor\_Voltage}(t)dt$$

The time-average motor voltage is the integral of the motor voltage over one period of the waveform divided by the period. For our waveform, the motor voltage is non-zero only from $t = 0$ s to $t = t_{on}$:

$$\left\langle V_{Motor\_Voltage} \right\rangle = \frac{1}{T_S} \int_0^{T_S} V_{Motor\_Voltage}(t)dt = \frac{1}{T_S}\left[\int_0^{t_{on}}(200\,volts)dt + \int_{t_{on}}^{T_S}(0\,volts)dt \right] = \frac{1}{T_S}\int_0^{t_{on}}(200\,volts)dt$$

This equation reduces to the integral of a constant (200 volts in this case), which is easily solved:

$$\langle V_{Motor\_Voltage} \rangle = (200 \text{ volts}) \frac{t_{on}}{T_S}$$

Since t_{on} can range from 0 to T_S, the average motor voltage can range from 0 to 200 volts. Note that the 200 volts was the voltage of our constant voltage source provided by batteries. In general, we can have higher or lower battery voltages and the general circuit is:

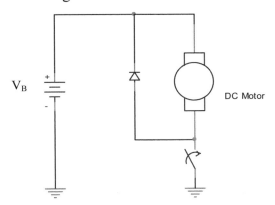

So, the average motor voltage is

$$\langle V_{Motor\_Voltage} \rangle = V_B \frac{t_{on}}{T_S}$$

By using pulse-width modulation, the average motor voltage can be varied from 0 to V_B by changing the on time, t_{on}. Since the motor responds to the average voltage, we can control the power to the motor by controlling the on time.

V.B. Switching Devices

Now that we understand how to control the power to the motor, how do we realize the circuit to accomplish the control? First, we will choose a switch. There are two devices that we can use, a power n-type MOSFET and an insulated gate bipolar transistor (IGBT). In general, MOSFETs can be switched at much higher frequencies than IGBTs. Typically, IGBTs are limited to a switching frequency of 20 kHz or lower, although this is strictly not the case. MOSFETs can have switching frequencies in the hundreds of kilohertz. Usually MOSFETs are used in lower-voltage and lower-power applications, and IGBTs are used in higher-voltage, higher-power applications, such as a 100 kW motor controller; however, the voltage and current capabilities of MOSFETs have been increasing, and IGBT manufacturers have been producing lower-voltage, lower-current devices. Thus, you should consider both devices when choosing a switch.

To turn on a power MOSFET, you apply a large voltage between the gate and source terminals (referred to as V_{GS}). By "large" we mean greater than the device's threshold voltage and less than 20 V, which is the typical maximum gate-source voltage allowed for power MOSFETs. Usually you turn on the MOSFET with as much gate drive as possible, which is typically 12 V or 15 V. When a MOSFET is an on switch, it looks like a small resistor R_{DSon}.

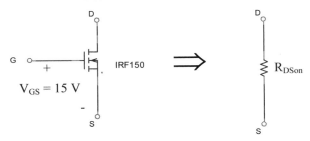

Typical values of R_{DSon} can be from a few hundred milliohms all the way down to 3 or 4 milliohms for the larger MOSFETs. You will need to refer to the datasheet to find the value of R_{DSon} for your particular device.

To turn off the MOSFET, we need to reduce V_{GS} well below the threshold, typically as close to zero as our drive circuit will provide. Note that we can also make V_{GS} negative to turn off the MOSFET. When we turn off the MOSFET, it looks like a very large resistance, which we will approximate as an open circuit:

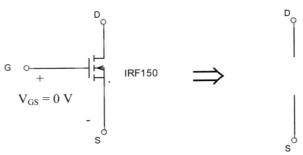

We will use a MOSFET in our example:

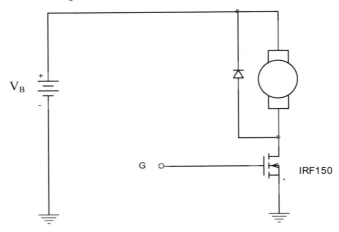

To pulse-width modulate the motor, we need only provide a 0 to 15 V PWM signal to the gate of the MOSFET. The voltage across the motor will be a PWM voltage from 0 to V_B.

An IGBT is used in a manner similar to the MOSFET. To turn on an IGBT, you apply a large voltage between the gate and emitter terminals (referred to as V_{GE}). By "large" we mean greater than the device's threshold voltage and less than 20 V, which is the typical maximum gate-source voltage allowed for power IGBTs. Usually you turn on the IGBT with as much gate drive as possible, which is typically 12 V or 15 V. An IGBT is actually a bipolar junction transistor (BJT) with a MOSFET input stage that supplies the base current to the BJT and provides high input impedance at the gate terminal. [9] When you turn on the IGBT, you are actually saturating the BJT. Thus, when the IGBT is on, it looks like a saturated BJT. The model for a saturated BJT is a small DC voltage source, V_{CEsat}. For BJTs, a typical saturation voltage is around 0.3 V. For an IGBT, the saturation voltage can be as high as 2 or 3 volts. You will need to refer to the datasheet to find the value of V_{CEsat} for your particular device. Thus, for IGBTs, when we apply a large gate-to-emitter voltage, the IGBT saturates and is an on switch. In this on state, we model the IGBT as a small DC voltage source:

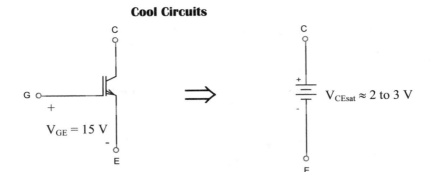

Since V_{CEsat} can be a large voltage compared to the voltage drop across R_{DSon} of a MOSFET, IGBTs are usually used in applications where V_{CEsat} corresponds to a small percentage of the supply voltage (V_B in our previous motor control circuit).

To turn off the IGBT, we need to reduce V_{GE} well below the threshold. You can turn off an IGBT by making V_{GE} equal to zero. For higher-power applications, however, IGBT manufacturers recommend that you turn off the IGBT with a negative V_{GE}, typically –10 V. When we turn off the IGBT, it looks like a very large resistance, which we will approximate as an open circuit:

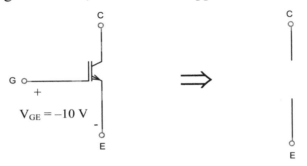

V.C. PWM Circuit

The next thing that we need to do is create the PWM signal to drive the gate of our MOSFET. This is realized by comparing a constant reference to a periodic ramp:

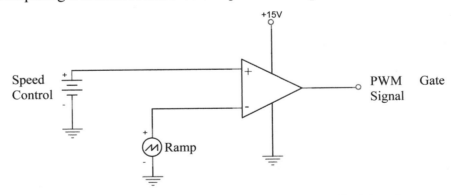

The speed control input is a constant or slowly varying signal controlled by the motor operator. The ramp is a periodic signal whose frequency determines the switching frequency. When the speed control voltage is greater than the ramp voltage, the comparator output will saturate at the positive supply, 15 V in this example. When the speed control voltage is less than the ramp voltage, the comparator output will saturate at the negative supply, 0 volts in this example. By "slowly varying," we mean that the control voltage changes at a much lower frequency than the ramp frequency. For our PWM waveforms shown earlier, the switching frequency was 1 kHz. To create this switching frequency with our comparator circuit, the ramp frequency should be 1 kHz. Thus, the speed control signal should change at a frequency much slower than 1 kHz.

To show how this circuit works, we will simulate the circuit using PSpice. The ramp signal will be a 0 to 5 V ramp at 1 kHz. We will use an LM324 op-amp because it can be used as a comparator and can switch at the relatively low frequency of 1 kHz. For the first simulation, we will set the speed control input to 2 volts:

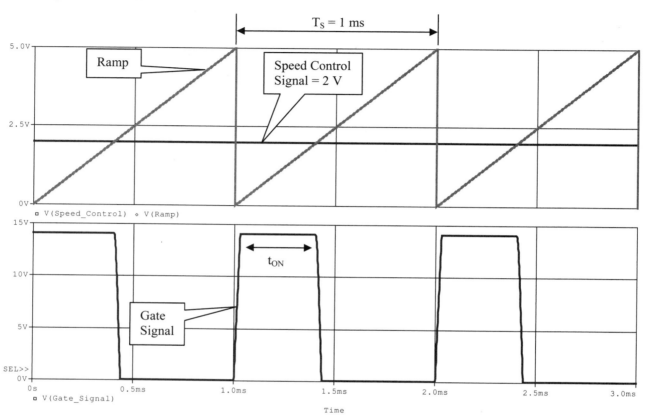

We see that when the speed control signal is greater than the ramp, the comparator output flips high. When the speed control input is less than the ramp, the comparator output goes low. The comparator output will control our MOSFET switch.

We can calculate t_{on} as:

$$t_{on} = \left(\frac{\text{Speed Control Voltage}}{\text{Ramp Peak Voltage}} \right)(\text{Switching Period}) = \frac{2\,\text{V}}{5\,\text{V}} \cdot 1\,\text{ms} = 400\,\mu\text{s}$$

The measured value of t_{on} from the waveforms above is about 380 μs. The simulated on time is slightly less than predicted because the op-amp takes time to switch from one state to the other. This limitation is due to the slew rate of the op-amp. [10] The LM324 is a fairly slow op-amp, and thus its output takes a fair amount of time to switch from low to high, and from high to low.

To show how we can control the on time, we will show the same waveforms with a control voltage of 4 V. For this input, the predicted on time is:

$$t_{on} = \left(\frac{\text{Speed Control Voltage}}{\text{Ramp Peak Voltage}} \right)(\text{Switching Period}) = \frac{4\,\text{V}}{5\,\text{V}} \cdot 1\,\text{ms} = 800\,\mu\text{s}$$

The waveforms are:

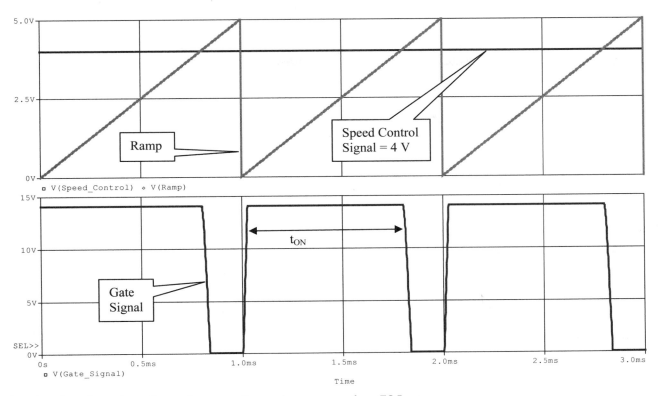

The simulated on time from the waveforms is measured as 785 μs.

We will now combine our PWM circuit with the switch and motor to show the complete motor control strategy:

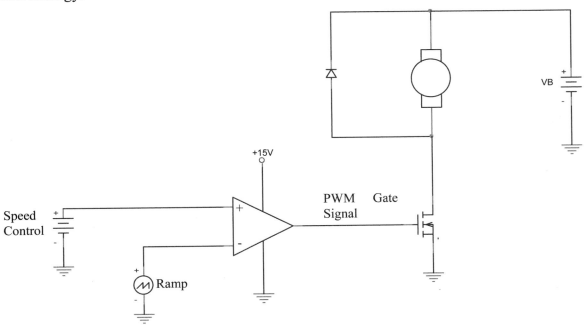

When the speed control signal is greater than the ramp, the switch will be on and the motor will see the full motor supply voltage V_B. When the speed control input is less than the ramp, the switch will be off and the motor will see zero volts. V_B can be a low-power supply for small motors, or a very-high-voltage, high-power supply for large motors. The control strategy works the same for all power levels. Example waveforms for $V_B = 200$ V are shown below:

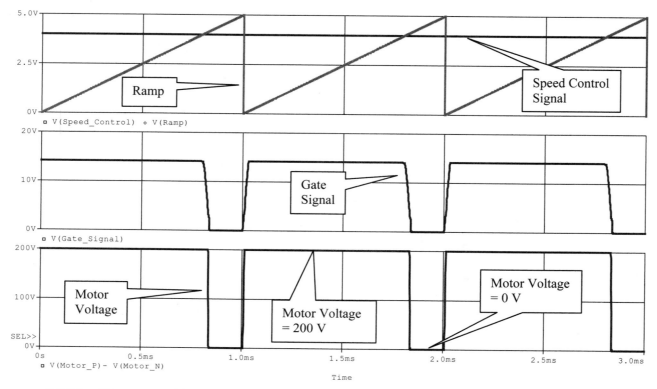

V.D. Ramp Generator

The last thing we need to do is to create the speed control and the ramp signals. We will generate the ramp signal using a UC3525A integrated circuit (IC). This chip is designed to control DC-DC power supplies such as the boost converter discussed in section IV. This chip uses PWM modulation to control DC supplies, and has a built-in ramp generator. It has many other functions that we will not use. We need only to choose a timing resistor and a capacitor to set the ramp frequency.

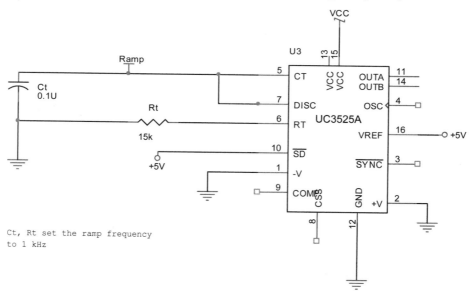

The datasheet for the UC3525A states that the oscillation frequency of the ramp is

$$F = \frac{1}{C_t \left(0.7R_t + 3R_D\right)}$$

R_D is called the dead-time resistor and determines the fall time of the ramp. In the circuit above, R_D is not shown and is zero to make the fall time as short as possible. Using the components shown on the schematic, $C_t = 0.1\ \mu F$ and $R_t = 15\ k\Omega$, we calculate the ramp frequency as:

$$F = \frac{1}{C_t(0.7R_t)} = \frac{1}{0.1\mu F(0.7 \cdot 15\,k\Omega)} = 952\,Hz$$

This is pretty close to 1 kHz using standard components.

Note that the UC3525A does not produce a 0 to 5 V ramp. Instead, it produces a ramp from 1 V to about 3.3 V. But this is the easiest way to produce a ramp, so we will live with it. Also note that the UC3525A has a 5 V reference, and we will use this reference for our speed control signal. The ramp waveform produced by the UC3525A is shown below:

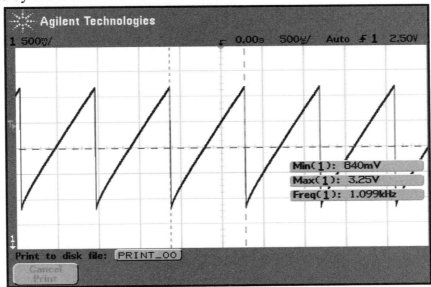

The ramp has a minimum value of 840 mV and a maximum value of 3.25 V. These values are typical for a UC3525A. Note that the min and max values of the ramp are not specified in the UC3535A datasheet, so the values you obtain could be quite different. However, I have measured the ramp produced by several different UC3525A chips, and the values of the ramp shown are typical. For my designs, I usually assume that the ramp is between 1 V and 3.3 V as an approximation. The frequency shown above is 1.099 kHz. This frequency is controlled by R_t and C_t, so you can set the frequency to anything you want, within the operating limits of the UC3525A. Also note that typical capacitors have a $\pm$ 20% tolerance, so your measured ramp frequency could be quite different from what you calculated.

The speed control signal is created by using a potentiometer, which was discussed in detail on page 6. We will use the 5 V reference of the UC3525A and the potentiometer to create a variable 0 to 5 V signal. The potentiometer is used as a variable voltage divider, and the division ratio is controlled by the user with a small knob. Remember that the ramp is a 1 V to 3.3 V signal. Because the speed control signal can go below the ramp signal, we guarantee that the motor will turn off when the speed pot is turned below a certain point. Because the speed control signal can be greater than the ramp signal, the motor can be full on.

A complete circuit for a low-power motor is shown below:

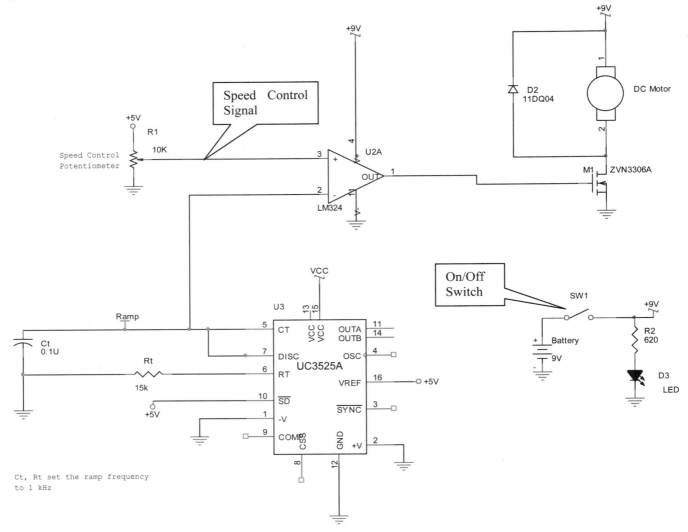

This circuit implements a small DC motor controller that runs from a 9 V battery. The unit is a little handheld device as shown below:

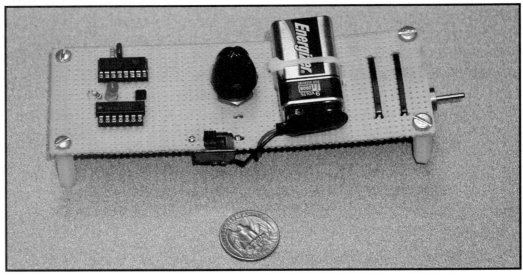

Although this demo circuit is a very-low-power circuit, the same principles can be used to control very large motors. The picture below on the left shows a motor controller designed to work with voltages in excess of 200 V and average currents as high at 600 A. This controller was designed for an electric race car that achieved speeds in excess of 120 mph.

A controller of this power does require a few more support circuits such as a gate driver, gate protection, and low-inductance layout techniques, but the underlying control principles are the same as discussed here.

The picture above on the right is of the Northern Arizona University Formula Lightning electric race car. This race car used a 200 V battery pack. During short bursts of acceleration, the pack and motor controller could supply 600 A for a few seconds. The car would cruise at a current of 300 A, which corresponds to 60 kW or 80 hp of power. The car had a limited range of about 8 miles because the batteries cannot supply this level of power for very long.

Cool Circuit VI
Electronic Hot Dog Cooker

Suppose you are having a Fourth of July barbecue and the main dish is hot dogs. You attempt to light your gas grill and find that you are out of gas. Because the Fourth of July is a national holiday, there are no propane dealers open to fill your tank. You run to the nearest convenience store to exchange your propane tank and find that all of the tanks are empty because everyone else in the country started their barbecue before you, ran out of gas, and then exchanged their tank at the convenience store. Your guests are getting hungry and starting to tell jokes about electrical engineers. As a last resort, you attempt to microwave the hot dogs, only to realize that your microwave broke several years ago, and you have not yet had a chance to fix it. Your guests are now threatening to leave your barbecue and crash the one being held by a mechanical engineer down the road. (At least a mechanical engineer understands heat transfer and how to cook hot dogs.) What do you do?

Suddenly, you remember a reference a professor of yours made twenty years ago about a hot dog being a conductor, and a "quickie hot dog cooker."[1] You find an extension cord, cut off the female end, strip the two wires (ignoring the ground wire), and connect the two wires to two ¼-20 stainless steel bolts (because you wish to observe the restaurant regulations in your city). You machine down the end of the bolts to a sharp point that can skewer a hot dog. (Being an electrical engineer, you would probably use a file or sander to sharpen the bolt rather than a lathe, but we will skip the details here.) You place the hot dog on a non-conducting plate, tell your guests not to touch the hot dog or the stainless steel bolts, and plug in the extension cord. At first nothing happens. You watch and wait. Then you begin to hear something. It starts out very quietly and slowly grows. The hot dog is sizzling! Next, a bubble forms in the skin of the hot dog. The sizzling has become loud and clear. Suddenly, the bubble pops and steam spews out of the rupture. A fissure splits down the length of the hot dog and steam billows out. The hot dog splits completely open and its insides spew out. Suddenly, one of the stainless steel electrodes pops out of the hot dog and the show is over. You unplug the extension cord.

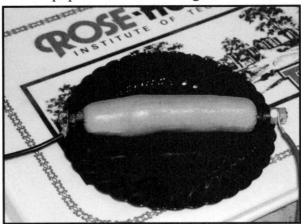 

---

[1] Thanks to Professor Mark N. Horenstein of Boston University for mentioning the quickie hot dog cooker those many years ago.

Your barbecue is saved. No one wants to eat the hot dog, but everyone wants to cook another. An autopsy of the hot dog reveals that the meat that was in contact with the stainless steel electrodes has turned black, but was not burned. It appears that a chemical reaction occurred at the electrodes. The photos below show pictures of the hot dog after it was cooked. One photo shows that the hot dog basically exploded. The second photo shows what the inside of the hot dog looks like. It appears that the hot dog is badly discolored where it came in contact with the electrodes.

A hot dog is a conductor. When you apply 115 VAC RMS across the hot dog, it will draw between 2 and 3 A RMS of current. Thus, the power absorbed by the hot dog is around 200 to 300 watts. The reason the hot dog split open and spewed its guts out was because this is too much power to evenly cook the hot dog. We need a method to regulate the power to the hot dog.

VI.A. Variac

A variac is a variable transformer. If you remember, a transformer will convert an AC voltage to a higher or lower voltage. Ideally, the power input to a transformer equals the power output, so if the output voltage is higher than the input voltage, the output current will be lower than the input current. A typical drawing for a transformer is shown below:

The drawing is meant to suggest two coils wound on the same core. The voltage and current relationship between the two windings depends on the number of turns in each winding. An important point we want to emphasize is that the transformer gives us electrical isolation between the input and output. There is no direct wire connection between the two sides of the transformer. Power is passed from one winding to another through the magnetic field in the core. In low-voltage supplies where the output of the transformer is at a much lower voltage than the input, one of the most important functions of the transformer is the isolation between the transformer's low-voltage output and the 115 VAC line voltage. The transformer reduces the chance of the user of the 12 V output coming in direct contact with either of the two wires of the 115 V line.

A variac has a slightly different symbol:

The symbol is reminiscent of a potentiometer. A way to look at a variac is that it is an inductive voltage divider where we change the division ratio just like we did with a potentiometer. (This is not strictly true because the output of a variac can be higher than the input, which is not possible with the voltage divider analogy.) A variac has a direct wire connection between its input and output terminals, so there is no electrical isolation. This means that it is possible for a user working with the output terminals of the variac to come in direct contact with the 115 V line. Thus, you should not use a variac when you want to prevent a direct connection between the output and the 115 V line.

For our purposes, a variac will give us an AC voltage from 0 to 115 VAC, and this will allow us to cook our hot dog at any power we deem appropriate. The circuit we will use is:

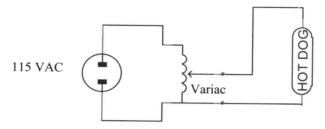

This circuit allows us to easily regulate the power to the hot dog, allowing us to slowly cook the hot dog to achieve peak flavor and appearance. (Ketchup and mustard are still recommended.) There are two major problems with this method. The first is that the hot dog is not electrically isolated from the 115 V line. If someone touches the hot dog while it is cooking, they could receive a 115 VAC shock even if the variac is set to low output voltages. The second problem is that variacs are fairly expensive, so this solution is not very economical.

VI.B. PWM Power Regulation

We will use pulse-width modulation (PWM) to regulate the power to the hot dog. We will use almost the same method we used for the DC motor controller of section V. You should read section V before continuing because it explains the theory of PWM control and the circuit used to implement it.

As a first attempt, we will directly cook the hot dog from the AC line using the circuit below:

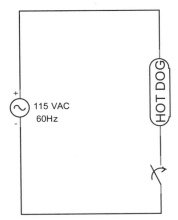

This circuit will work, but it requires a switch that can stop AC voltages and currents. When the switch is on, it must allow current to flow in both directions; when off, current must not flow in either direction. The switches discussed in section V.B are DC switches in the sense that they can only stop current from flowing in one direction. If you look at the symbol for power MOSFETs and IGBTs, you will notice a diode in parallel with the device. This diode is sometimes referred to as the anti-parallel diode or the body diode. [11] An example symbol is shown below:

The switch can stop current from flowing in the direction shown above. Current can never flow through the diode (unless you apply a voltage large enough to break it down) and the MOSFET can turn on and off, which will turn the current on and off. The switch cannot stop current from flowing in the direction shown below:

Even if the MOSFET is off, current can always flow through the diode, making the MOSFET irrelevant. Thus, we can only use a power MOSFET or IGBT to stop current from flowing in one direction.

Using two MOSFETS, we can make a switch that stops current from flowing in both directions:

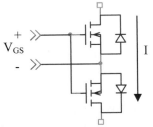

Note that the two gates and two sources of the MOSFETs are tied together. The top switch can turn current on and off in the direction shown above because the top body diode is reverse biased for this current. The bottom MOSFET has no effect on the current because current can flow through the diode whether the bottom MOSFET is on or off. When current flows in the opposite direction, the bottom switch controls the current:

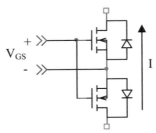

The bottom body diode is reverse biased for this current, so the bottom MOSFET can turn this current on and off. The top body diode passes current in the direction shown, so the top MOSFET has no effect on this current. With the arrangement shown, we can turn current on and off in both directions, and the controlling voltage is shown as V_{GS}. The same requirements apply to turning this switch on and off. To turn it on, you need V_{GS} much greater than the MOSFET's threshold voltage, typically 15 V. To turn off the switch, you need V_{GS} much less than the threshold, typically zero volts.

The one disadvantage of this circuit, and the reason we will not use it, is that the gate drive circuitry can be more complicated depending on how you use the switch. Since the two sources are connected together, the sources may not be grounded in some applications, and the gate-source voltage will not have a ground reference. This requires a more complicated gate drive circuit, which we do not want to use here.

For simulation purposes, we can easily generate a floating gate drive signal (one that is not grounded) and observe the waveforms that we would see if we used this circuit. We will use the circuitry covered in section V using our new switch:

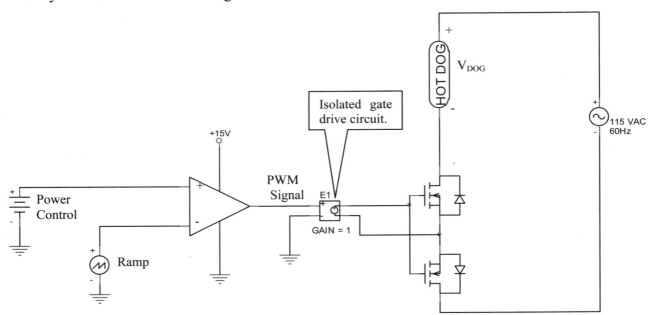

Note that the sources of the two MOSFETs are not connected to ground. To turn the MOSFETs on and off, we must apply the correct voltage between the gate and source of the MOSFETs. The gate-source voltage is not the same voltage as the output of the comparator because the comparator output is referenced to ground and the gate-source voltage is not. To drive the MOSFETs we are using a voltage-controlled voltage source E1. The output of E1 is the gain times its input. The gain is 1 and the input to E1 is the comparator output. Thus, the output of E1 is equal to the comparator output. However, the output of E1 is directly connected between the gates and sources of the MOSFETs. Thus, the PWM signal at the output of the comparator is directly applied to the gate-source voltage of the MOSFETs, even though the gate-source voltage is not referenced to ground. In practice, this type of gate drive can be accomplished with an optically isolated gate drive integrated circuit and an isolated power supply to power the gate drive circuit. In PSpice simulations, however, isolated gate drive is easily accomplished with a single part, E1.

We will run the same simulations as we did for the DC motor controller and look at the voltage across the hot dog. The only difference will be that the voltage across the hot dog will have a sinusoidal envelope.

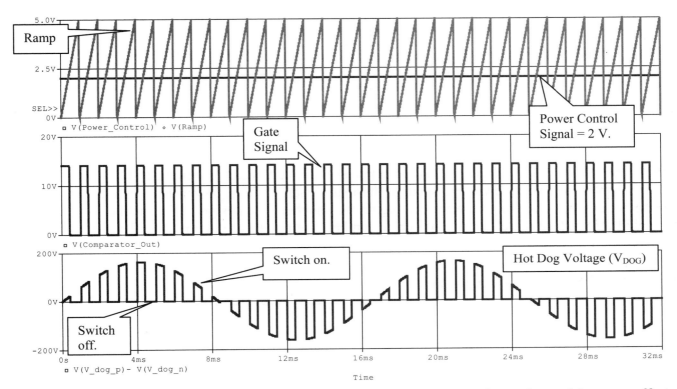

Thus, we can accomplish pulse-width modulation of a sine wave. It just takes a bit more effort to generate the gate drive circuitry and the AC switch.

There is another problem that we have not addressed. When we used PWM modulation to control the DC motor, we added a freewheeling diode in parallel with the motor to eliminate voltage spikes that would occur when we turn off the current through an inductive load, such as a motor. We need similar protection for our hot dog cooker. You might be tempted to say that the hot dog is a purely resistive circuit element and therefore the current through the hot dog can change instantaneously, and you may be correct. (When you look at the Nutrition Facts label for a hot dog, it does not say anything about a hot dog being inductive. This is the closest we can come to a datasheet for a hot dog.) However, we will connect long wires to the hot dog, and these wires will have some inductance. [12] Thus, as a good circuit designer, we should prevent any voltage spikes that might occur due to the inductance of these wires. In the motor controller, which used a positive voltage, we could use a freewheeling diode. We cannot use that same diode here because we are using an AC voltage across the hot dog. To suppress the voltage spikes we need to use back-to-back Zener diodes, a bidirectional Transient Voltage Suppressor (TVS), or a Metal Oxide Varistor (MOV) as shown in the circuit below:

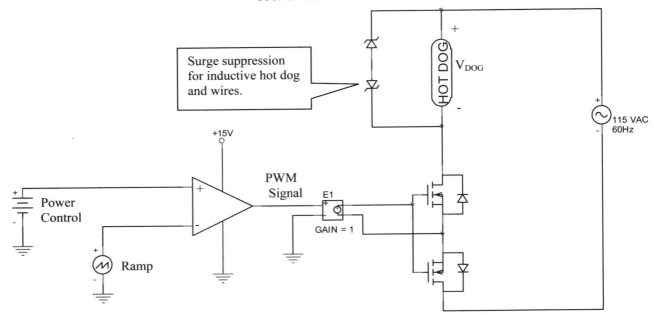

In the world of power electronics, this surge suppression circuit is called a snubber. [13] Each time the snubber absorbs an inductive voltage spike, it dissipates energy. Thus, power that was intended to flow into the hot dog will instead go toward heating up the snubber. The design of the snubber is non-trivial, and it will also reduce the efficiency of our design because not all of the power is absorbed by the hot dog.

We can avoid the problem of using a bidirectional switch and the snubber if we rectify the 115 VAC input waveform to make it positive:

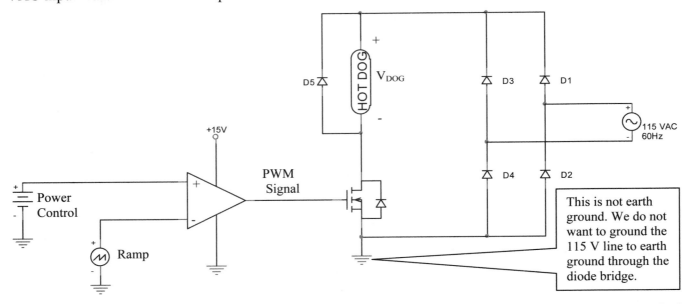

The voltage across the hot dog (V_{DOG}) will always be positive due to the full-wave rectifier composed of D_1-D_4. This allows us to use a single MOSFET as the switch, and we can use a freewheeling diode (D_5) rather than a more complicated snubber. Simulated waveforms for this circuit are shown below:

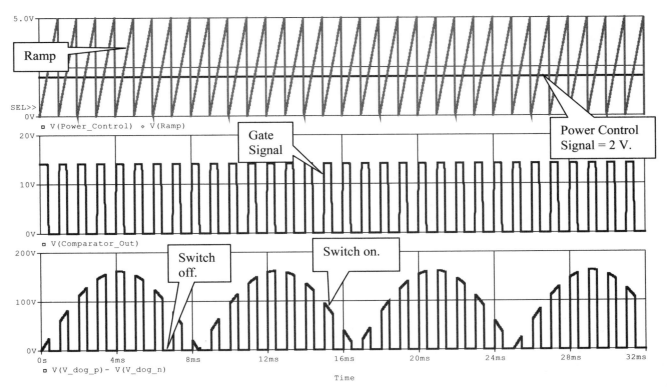

The power to the hot dog is controlled by changing the power control signal in the same way that we changed the speed control signal in the DC motor controller. To increase the power, we increase the DC voltage of the power control signal. This will increase the on time of the switch and the power to the hot dog. An example with the power control signal set to 4 V is:

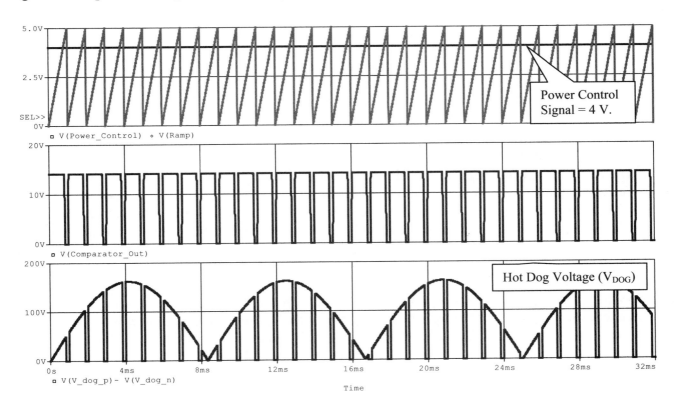

We see that the voltage across the hot dog is nearly a complete sine wave, which delivers close to full power to the hot dog. To reduce the power to the hot dog, we reduce the DC voltage of the power control signal, which will reduce the on time and the power to the hot dog.

VI.C. Complete Electronic Hot Dog Cooker Circuit

The complete circuit for the hot dog cooker is:

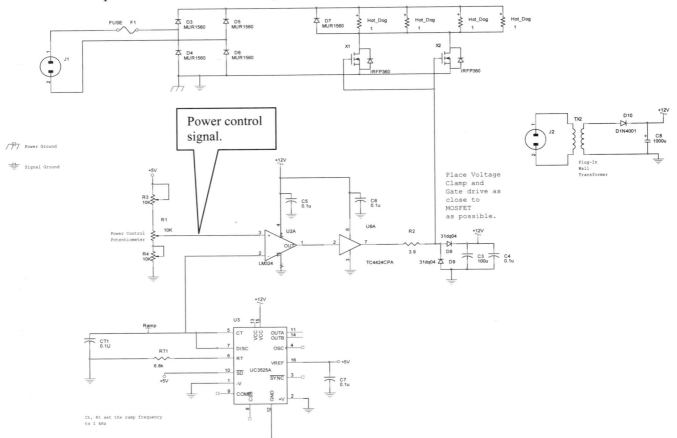

The circuit is nearly identical to the DC motor controller with a few notable exceptions. Since we are running our circuit from the 115 VAC line, we need to create a +12 V DC supply to run our circuit. This is done with a plug-in wall transformer, a diode (D_{10}), and a capacitor (C_8). Off-line DC supplies were discussed in section I.B, so the supply will not be discussed here.

The circuit has been designed to cook four hot dogs (although I have never tested it with four hot dogs to see if it can handle the power) and uses two large power MOSFETS in parallel as the switch (X_1 and X_2). You can easily parallel MOSFETs to get a higher-current switch. Because we are using large MOSFETs, and we are building a high-voltage and high-power circuit, a little more care is needed in the gate drive circuitry. A typical op-amp or comparator output can source and sink only a few milliamperes of current. Between the gate and source of a MOSFET is a built-in capacitor, and we need to charge and discharge this capacitance quickly if we want to turn the MOSFET on and off quickly. To charge and discharge the gate-source capacitance we will use a TC4424 gate driver, which is capable of driving capacitive loads with a peak current of 3 A. This will turn the MOSFET switches on and off much faster than the comparator output could, and will lead to more efficient operation of the MOSFET switches. R_2 was chosen as 3.9 Ω to limit the gate driver current to 3 A peak (12 V/3.9 $\Omega \approx$ 3 A).

The circuit composed of D_8, D_9, C_3, and C_4 is a clamping circuit that protects the gates of the MOSFET from receiving voltages greater than their maximum rated voltage of $\pm$ 20 V. C_3 and C_4 charge up to 12 V and provide a local +12 V supply that is physically close to D_8 and D_9. Think of C_3 and C_4 as a 12 V battery for analysis purposes. D_8 and D_9 are Schottky diodes that turn on and off very quickly and have a turn-on voltage of about 0.3 V. D_8 prevents the gate voltage from becoming larger than about 12.3 V while D_9 prevents the gate voltage from going below –0.3 V. In this high-voltage switching application, the drain of MOSFETs will be switching between 0 and 162 V. This high-voltage square wave on the drain can be fed back to the gate through the built-in gate-drain capacitance of the MOSFET. To protect the gate from high-voltage pulses, we add the clamp composed of D_8, D_9, C_3, and C_4.

One last difference between this circuit and the DC motor controller is the addition of the potentiometers R_3 and R_4. These are two trimmer potentiometers that we set and then never change. If you remember from the motor controller, the ramp produced by the UC3525A goes from about 1 V to about 3.3 V. Without R_3 and R_4, the power control signal would range from 0 to 5 V, and a large portion of its range would have no effect on the power we deliver to the hot dog. We added R_3 and R_4 so that the minimum value of the power control signal is just less than 1 V and the maximum value is just greater than 3 V. This allows the power control pot to control the power through its full range of movement. R_3 and R_4 are wired as variable resistors, so we can tailor the power control signal to the characteristics of the ramp.

VI.D. Electrode Design

One of the more difficult aspects of this design was the electrodes that we plug into the ends of the hot dog. The electrodes are shown below:

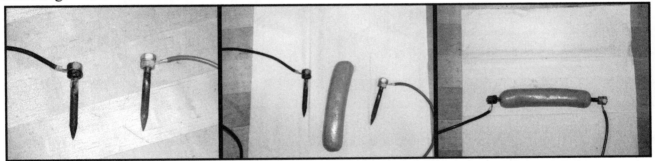

We used stainless steel ¼-20 bolts, removed the hex head of the bolts, and then sharpened the non-threaded end into a point. Stainless steel was used so that there was less chance of the metal reacting with the hot dog, and for local dining ordinances, which require the use of stainless steel cooking utensils. The threaded end of the bolt was left intact so that we could use a crimp lug to attach the wire to the bolt. We did not want to use solder to connect the wire to the electrode because solder contains lead, which would migrate into the hot dog as it cooked.

So, why did we choose a rather large diameter ¼-20 bolt? The surface area of the electrode is very important. First, we need good electrical contact between the hot dog and the electrode. More surface area gives us better electrical contact. Also, when the hot dog is uncooked, it contains a lot of liquid that gives us a good electrical connection. As the hot dog cooks, it dries out around the electrodes increasing the resistance of the hot dog in those areas. The increased resistance reduces the cooking power. If the electrode surface area is too small, it becomes hard to get electrical power into the hot dog without repositioning the electrodes as it cooks. This problem is avoided by using large-area electrodes such as a ¼-20 bolt.

VI.E. Waveforms and Cooked Hot Dogs

The circuit works very well. Some waveforms of the voltage across the hot dog going from low power to high power are shown below:

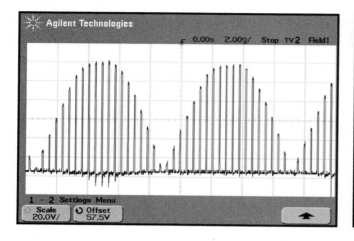

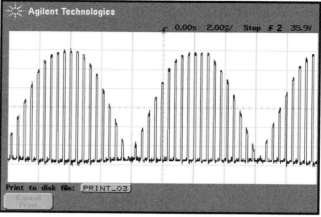

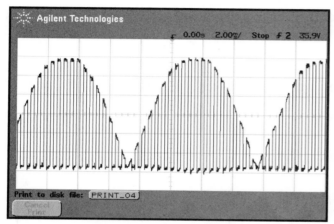

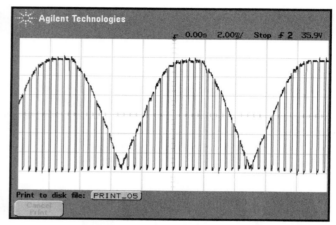

An enlarged version of a low-power and a high-power waveform is shown below. These waveforms show the difference in the pulse width a little better:

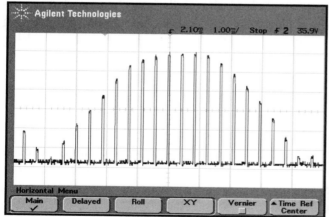

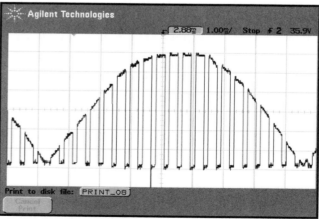

When we look inside a cooked hot dog, we notice that the color of the meat at the two electrodes is different. With the AC-cooked hot dog, the ends at both electrodes were black. With the DC-cooked hot dog, we notice that one contact area is black and the other is a bright rust-colored red. We don't know why, but it is interesting to note the difference. (This could be a good topic for a government-funded research contract!) The PWM hot dog cooker is shown below. The circuit shown uses only a single MOSFET rather than two MOSFETs in parallel. Thus, the circuit will not be able to cook as many hot dogs as if we had used two MOSFETs in parallel. I used only a single MOSFET because I made only a single pair of electrodes, so I can only cook one hot dog at a time regardless of how much power the circuit can handle.

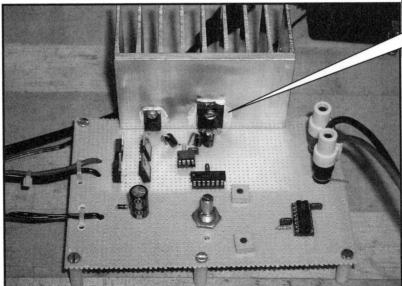

Single power MOSFET.

Cool Circuit VII
Magic Feedback Audio Amplifier

In this example we will discuss a push-pull amplifier with crossover distortion and how to magically reduce the distortion using feedback. We will first discuss the motivation for using a push-pull power amplifier. Next, we will discuss how feedback can be used to reduce distortion. Finally we will use feedback with the push-pull amplifier to create a high-power audio amplifier with a small amount of distortion. The amplifier we will create can source about 50 W peak to a 4 Ω load.

VII.A. Background

VII.A.1. Power Amplifiers

Suppose we have an audio signal source such as a CD player, and we want to amplify the signal and then deliver that amplified signal to a 4 Ω load. We first need to amplify the signal using an op-amp or a discrete amplifier such as a common-emitter amplifier. The problem with these types of amplifiers is that they cannot drive large loads; connecting a large load to an op-amp will cause it to saturate. Connecting a large load to a common-emitter or common-source amplifier will drastically reduce its gain. These two topologies can give us voltage gain, but they cannot drive large loads. To fix this problem, we need to follow our voltage gain stage with a current amplifier that has a high current gain. An example of such a stage is an emitter-follower amplifier. If you remember, an emitter-follower amplifier has a small-signal voltage gain close to one, and it has a large current gain, so the output current is much larger than the input current. A typical circuit is shown below:

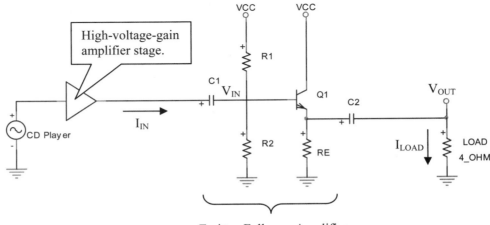

Emitter-Follower Amplifier

The high-voltage-gain amplifier is shown schematically as the triangular symbol. We assume that the signal coming out of this amplifier has the desired voltage swing, and we are using the emitter-follower to deliver that large voltage swing to the 4 Ω load. If you covered the emitter-follower in class, you will recall that V_{OUT}/V_{IN} can be close to 1, and the load current is much greater than the input current, $I_{LOAD}/I_{IN} > 1$.

If you proceed blindly by biasing the amplifier and then doing the small-signal analysis, you may think that you have designed an amplifier that can drive a large load. Note that by "large load," we mean a large load current or a small load resistance. For this example, we are calling the 4 Ω load a large load. As an example, we will use the NPN emitter-follower amplifier shown below:

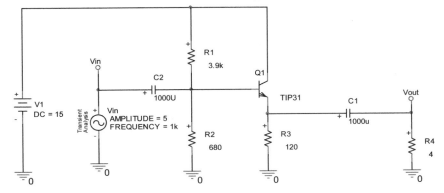

When you test the amplifier in the lab or with a simulation program, you will see the waveform below:

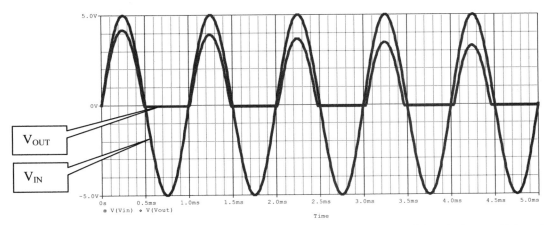

We see that the NPN emitter-follower amplifier can source the positive halves of the sine wave, but not the negative halves. During the positive portion of the sine wave, the amplifier has a gain of about 0.85. In our introductory electronics courses, we learned to do the small signal-analysis on this amplifier and we expect it to be able to supply a sine wave to the load; thus, the above result surprises us. After beating our brains for a while, we realize that for large load currents and large input voltages, this amplifier is not a small-signal amplifier. Thus, the small-signal analysis is not valid. For smaller load currents and smaller input voltages, we would see a sine wave at the output, but for our design we want to drive a large voltage swing into a large load. It turns out that for large voltage and current swings, when the input voltage becomes a large enough negative voltage, we will actually turn off the base-emitter junction of the device and the output current clips. We conclude that the NPN emitter amplifier can source only the positive halves of the sine wave.

If we examine a PNP emitter-follower, such as the one shown below, we will see that it can only source the negative portions of a sine wave:

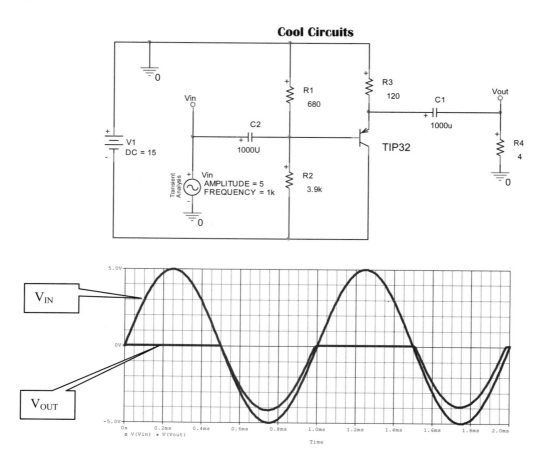

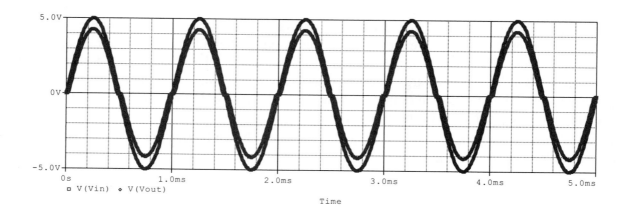

The obvious solution is to combine the NPN and PNP emitter-follower amplifiers into a single circuit called a push-pull amplifier. Its simplest realization has no biasing and is shown below:

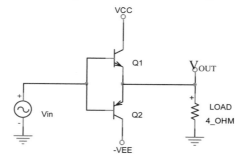

Example input and output waveforms for this amplifier are:

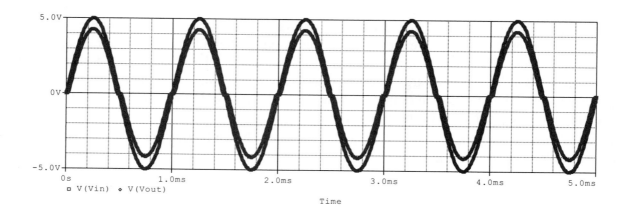

We see that the amplifier can now supply both the positive and negative halves of the waveform. We also, however, see a problem when the input signal crosses zero. Since this amplifier has no biasing, the 0.7 volts needed to turn on the base-emitter junction of the transistors must be supplied by the input voltage source. Thus, when $|V_{IN}| < 0.7\text{ V}$, both transistors are off and the output voltage will be zero.

This type of distortion is called crossover distortion. If we use a small input voltage, the crossover distortion will become more obvious. The waveforms below are for a 1 V amplitude input:

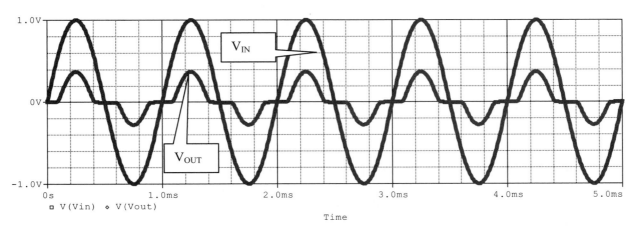

There are several ways to eliminate the crossover distortion. The most obvious is to add 0.7 V sources to provide the turn-on voltage for the transistors. In the circuit below, we have used diodes to create the 0.7 V sources:

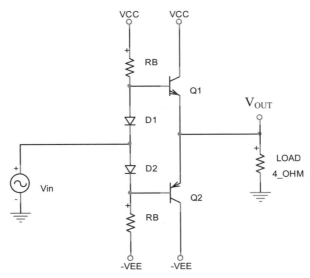

This amplifier fixes the crossover distortion, but it creates other problems such as a large amount of DC current flowing through Q1 and Q2 when the input signal is zero. This is a DC current that flows from VCC to −VEE and is sometimes referred to as the circulation current. This current is wasted power and can be quite large if there is a bad match between the diodes and the transistors. Worse yet, this wasted power heats up Q1 and Q2 and causes the amplifier to become thermally unstable, which can result in the destruction of Q1 and Q2. We can fix this problem by thermally coupling the bias diodes to Q1 and Q2 and by adding emitter resistors. Thus, we do need to improve the push-pull amplifier shown above, but we will not show how to do it here. Here, we are going to eliminate the crossover distortion using feedback.

We will make one last change to our push-pull amplifier. We will be driving a large current to our load, and thus the transistors must carry a large current and must be power transistors. Since power transistors are designed to carry large currents, they usually have a small current gain (H_{FE}) as compared to a small-signal, low-power transistor. A typical power transistor will have a current gain between 5 and 25. The consequence for our amplifier is that the input source must supply a large current. To reduce the input current, we need to create a power transistor with a large current gain. This can be done by using two transistors connected in what is called the Darlington configuration: [14]

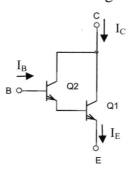

Q_1 carries the majority of the current and is a power transistor with a low current gain. Q_2 provides the base current for Q_1 and is a lower-power device because the base current of Q_1 is much smaller than the collector current of Q_1. Thus, Q_2 can be a small-signal device with a high current gain. As shown, the device can be thought of as a single three-terminal device. It requires two base-emitter voltage drops to turn on, but the overall current gain of the Darlington device is approximately the product of the individual current gains of each transistor:

$$I_C = H_{FE}I_B$$
$$I_E = (H_{FE}+1)I_B$$
$$H_{FE} = H_{FE1}H_{FE2} + H_{FE1} + H_{FE2} \approx H_{FE1}H_{FE2}$$

Thus, we can easily create a high-current-gain transistor using the Darlington connection. For a TIP102 Darlington power transistor, the minimum HFE is specified as 1000 at a collector current of 3A, which can be compared to a TIP31 power transistor that has a minimum HFE of 10 at a collector current of 3A.

Using Darlington transistors, our push-pull amplifier with crossover distortion is:

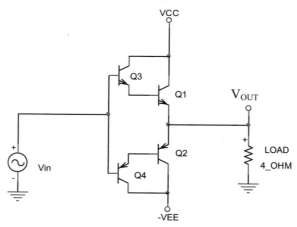

Note that this Darlington push-pull amplifier has more crossover distortion than the non-Darlington version because the input source must supply two base-emitter voltage drops to turn on the transistors. This is the version of the amplifier we will use in the following sections.

VII.A.2. High-Current Operational Amplifier

Operational amplifiers (op-amps) are very easy to use, but they do have limitations. One of the most severe limitations is that they can only supply a few milliamperes of current to a load. As an example, the non-inverting amplifier shown below will work as designed as long as the load current is only a few milliamperes or less:

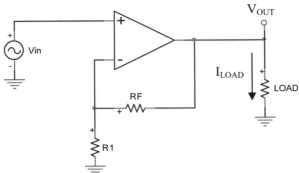

For load resistances in the kilohm range, the op-amp will only have to supply a few milliamperes to the load, and the op-amp will work properly. However, if we use a large load, say 4 Ω, the op-amp cannot supply the large currents that would be drawn by the load, and the output of the op-amp current limits.

We can create a high-current op-amp by cascading the op-amp and our push-pull amplifier. An example of a high-current op-amp is:

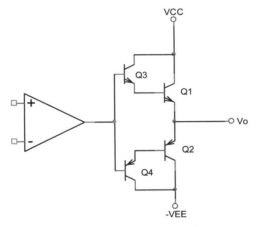

The op-amp has a large open-loop gain, typically close to 10^6. The push-pull amplifier has a gain close to 1, so the cascaded gain is still very large. The push-pull amplifier can drive a large load, but it does make the cascaded amplifier nonlinear by adding crossover distortion. We have created a high-voltage-gain, high-current amplifier with distortion.

VII.A.3. Feedback

We would like to see how feedback can be used to reduce the nonlinear properties of an amplifier. Suppose we have the system below:

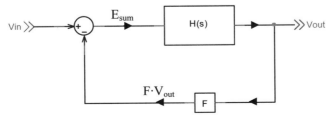

This is a feedback block diagram for a general system H(s). For this example, the system H(s) will be an amplifier with a large gain. We will not be looking at frequency response, so we will just use H(s) = G, where G is the open-loop gain of our amplifier.

We will now do the analysis for this system. The input to our feedback block F is the output voltage V_{OUT}. The output of the feedback block is F times its input, or $F \cdot V_{out}$. This voltage is the input to the negative terminal of the summing junction, so $E_{SUM} = V_{IN} - F \cdot V_{out}$. E_{SUM} is the input to our system G. The output of the system is G times its input, so

$$V_{OUT} = G \cdot E_{SUM}$$
$$V_{OUT} = G \cdot \left(V_{IN} - F \cdot V_{OUT} \right)$$

Solving the second equation for V_{OUT}/V_{IN} yields:

$$\frac{V_{OUT}}{V_{IN}} = \frac{G}{1 + FG}$$

For us, G will be the gain of an operational amplifier, which is very large, and F is typically created by using resistor networks. If FG >> 1, then 1 + FG ≈ FG and our equation reduces to:

$$\frac{V_{OUT}}{V_{IN}} = \frac{G}{1 + FG} \approx \frac{G}{FG} = \frac{1}{F}$$

Thus, for large open-loop gain G, the gain with feedback V_{OUT}/V_{IN} is independent of the properties of the amplifier G and dependent only on the resistor feedback network F. This is very useful result because G can vary by an order of magnitude or more, can be temperature dependent, or, as for our example, can be nonlinear. When we use feedback, however, all of these undesirable properties of the amplifier are greatly reduced, and the gain depends on the feedback network F, which is composed of resistors that can have a very tight tolerance, are linear, and are not nearly as temperature dependent as active devices inside an amplifier.

We will use this property of feedback to eliminate the crossover distortion of our push-pull amplifier.

VII.B. Push-Pull Amplifier with Feedback

We will now add feedback to our high-current op-amp to reduce the nonlinearities (distortion) introduced by the push-pull amplifier. For this example, we will use unity feedback (F = 1) where the output is tied directly back to the negative terminal. There are two ways we could connect the feedback. The circuit below has the feedback connected around the op-amp only:

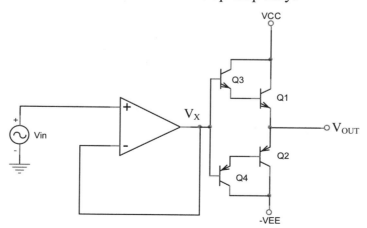

In this topology, the feedback will reduce the nonlinearities introduced by the op-amp because the op-amp is enclosed within the feedback loop; however, the crossover distortion introduced by the push-pull amplifier will not be affected. In this circuit, the op-amp is wired as a buffer with a gain of 1, $V_X/V_{IN} =$

1. V_X will follow V_{IN} with very high fidelity, but V_{OUT} will contain a large amount of distortion. A simulation showing V_{IN}, V_X, and V_{OUT} for a 4 Ω load is shown below:

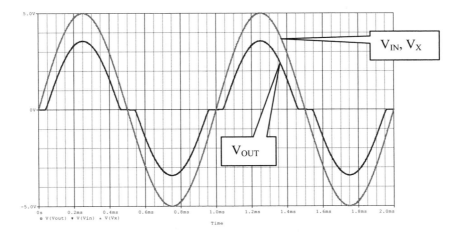

The input for our simulation is a 5 V amplitude, 1 kHz sine wave. From the simulation, we see that V_X is equal to V_{IN}, and both appear to be pure sine waves. The output, as expected, contains crossover distortion.

The output signal of our amplifier is obviously distorted. You can think of distortion as the frequencies contained in the output signal that are not contained in the input signal. In a distortion-free amplifier, if the input contains a single frequency, the output will also contain only a single frequency, the same as the input. For our amplifier, the input was a single-frequency sine wave at 1 kHz. To see the frequencies contained in the output, we can use PSpice to view the Fourier components of the output signals. We can view the components graphically or as a text file. Below are the Fourier spectra of V_{IN}, V_X, and V_{OUT}. The top trace is V_{IN} and V_X, and the bottom trace is V_{OUT}:

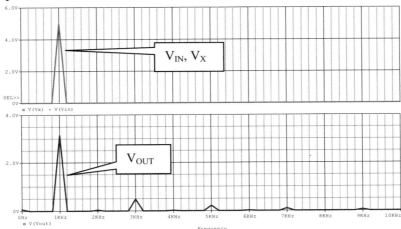

We see that the V_{IN} and V_X traces contain a single peak, indicating that the signals contain a single frequency, in this case 1 kHz. The output spectrum (V_{OUT}) contains several peaks, indicating that the output contains several frequencies; in this case the most obvious frequencies are 1 kHz, 3 kHz, and 5 kHz. Since the output contains frequencies not contained in the input, the amplifier adds distortion to the signal. We can get similar information in tabular form, with numerical values. Below is the table for the output signal V_{OUT}:

```
FOURIER COMPONENTS OF TRANSIENT RESPONSE V(VOUT)
DC COMPONENT =   5.861058E-02
```

HARMONIC NO	FREQUENCY (HZ)	FOURIER COMPONENT	NORMALIZED COMPONENT	PHASE (DEG)	NORMALIZED PHASE (DEG)

1	1.000E+03	3.148E+00	1.000E+00	-8.192E-02	0.000E+00
2	2.000E+03	2.744E-02	8.716E-03	-9.049E+01	-9.033E+01
3	3.000E+03	4.905E-01	1.558E-01	1.799E+02	1.801E+02
4	4.000E+03	1.885E-02	5.987E-03	-9.083E+01	-9.050E+01
5	5.000E+03	2.243E-01	7.125E-02	1.797E+02	1.801E+02
6	6.000E+03	1.233E-02	3.917E-03	-9.098E+01	-9.049E+01
7	7.000E+03	1.070E-01	3.398E-02	1.793E+02	1.798E+02
8	8.000E+03	6.935E-03	2.203E-03	-9.113E+01	-9.047E+01
9	9.000E+03	4.408E-02	1.400E-02	1.785E+02	1.792E+02
10	1.000E+04	2.552E-03	8.105E-04	-9.131E+01	-9.049E+01

TOTAL HARMONIC DISTORTION =   1.756004E+01 PERCENT

The tabular data tells us the amplitude and frequency of each frequency component contained in the output. The second column is the frequency of each component in hertz. The third column is the amplitude of the frequency component. We see that the output contains frequencies from 1 to 10 kHz. The largest frequency component is at a frequency of 1 kHz and has an amplitude of 3.148 volts. For a distortion-free amplifier, the output should contain only a single frequency—the same frequency as the input. We see, however, that the output contains several frequencies. For example, the output contains a component at 3 kHz that has an amplitude of 0.4905 V. The last line shown also tells us the total harmonic distortion of the amplifier, in this case 17.56%, which is fairly high because of the crossover distortion. Most of the distortion is added by the push-pull amplifier.

A better way to connect the feedback is:

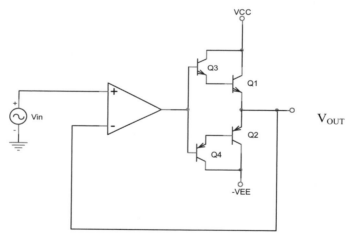

In this topology, the feedback encloses the op-amp and the push-pull amplifier, reducing the distortion introduced by both. A simulation showing V_{IN} and V_{OUT} for a 4 Ω load is shown below:

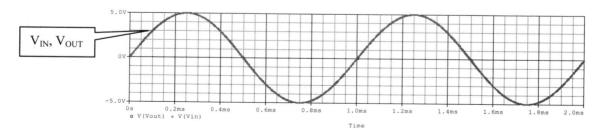

The feedback has reduced the distortion to the point that we cannot tell the difference between the input and the output. Two traces are shown on the plot, but they match each other.

To the naked eye, the above traces look so close to one another that we might assume that the amplifier is distortion free. A better way to determine the amount of distortion is to display the frequency spectrum of the input and output signals:

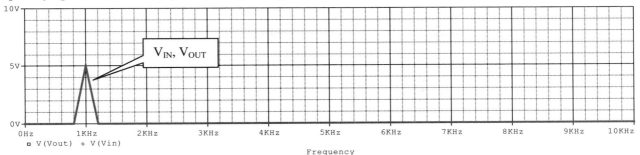

Both the input and the output contain a single peak at 1 kHz. Other peaks may be present in the output, but they are too small to see on this plot. To get a more accurate estimation of the frequency components of the output, we need to look at the text output of the Fourier analysis for this amplifier:

```
FOURIER COMPONENTS OF TRANSIENT RESPONSE V(VOUT)
DC COMPONENT =    1.154523E-05
```

HARMONIC NO	FREQUENCY (HZ)	FOURIER COMPONENT	NORMALIZED COMPONENT	PHASE (DEG)	NORMALIZED PHASE (DEG)
1	1.000E+03	5.000E+00	1.000E+00	-9.197E-02	0.000E+00
2	2.000E+03	5.382E-05	1.077E-05	-2.442E+01	-2.423E+01
3	3.000E+03	2.843E-03	5.686E-04	-9.414E+01	-9.386E+01
4	4.000E+03	2.108E-05	4.216E-06	-8.018E+01	-7.981E+01
5	5.000E+03	2.804E-03	5.608E-04	-9.693E+01	-9.647E+01
6	6.000E+03	2.047E-05	4.094E-06	-9.114E+01	-9.058E+01
7	7.000E+03	2.775E-03	5.550E-04	-9.978E+01	-9.914E+01
8	8.000E+03	2.024E-05	4.048E-06	-1.026E+02	-1.018E+02
9	9.000E+03	2.752E-03	5.503E-04	-1.027E+02	-1.018E+02
10	1.000E+04	2.065E-05	4.130E-06	-1.133E+02	-1.123E+02

TOTAL HARMONIC DISTORTION =   1.117570E-01 PERCENT

We see that the output does contain frequencies not contained in the input, so the amplifier does add distortion. Remember that the input was a single frequency at 1 kHz. The largest frequency component in the output is the same frequency as the input, and the other frequency components are very small. The frequency component at 1 kHz has an amplitude of 5 V. The largest frequency component not contained in the input signal is the third harmonic at 3 kHz, which has an amplitude of 2.843 mV. Thus, this amplifier is not distortion free, but the size of the components is much smaller than before. The last line tells us the total harmonic distortion of this amplifier is 0.112%, which is much smaller than the 17.56% for the previous feedback connection.

We see that the feedback greatly reduces the distortion added by the push-pull amplifier. This technique of using feedback to reduce distortion can be used with any amplifier where the gain is high and you can enclose the amplifier in a negative feedback loop. Had we used a push-pull amplifier that did not have crossover distortion to begin with, we could have reduced the distortion of our amplifier by at least another order of magnitude. We chose this example so that we can visibly see the results using an oscilloscope.

To demonstrate this principle in the lab, we use the circuit shown below:

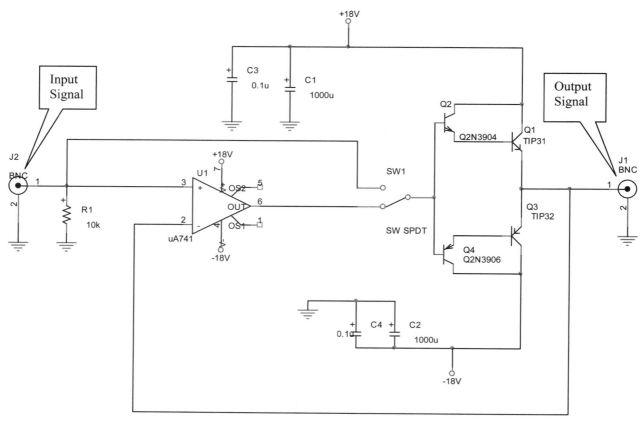

The input signal (J2) will come from a signal generator, or possibly from another op-amp stage that is used to amplify a signal coming from a CD player or some other audio source. The output (J1) can be connected to a 4 Ω speaker. The toggle switch (SW1) allows us to open or close the feedback loop. When SW1 is in the down position (as shown in the schematic) the feedback loop is closed, and both the op-amp and push-pull amplifier are contained within the feedback loop; we will have a low-distortion amplifier. When SW1 is in the up position, the push-pull amplifier is directly connected to the signal source, and the op-amp is removed from the circuit. In the up position, there is no feedback and we will see crossover distortion in the output.

Below are scope traces of the input and output waveforms when the switch is in the up position:

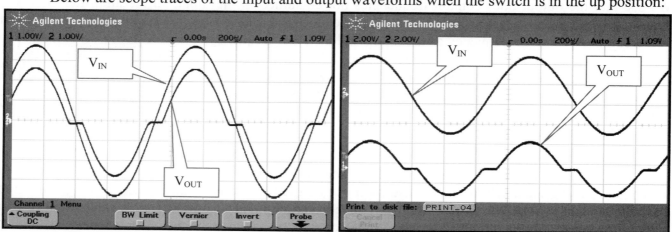

The output waveform obviously contains crossover distortion. The fast Fourier transform (FFT) of the output waveform is:

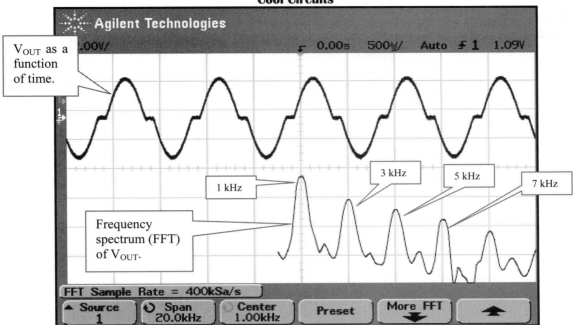

The scale of the FFT is 2 kHz per major division in the x-direction. Thus, the output contains large frequency components at 1 kHz, 3 kHz, 5 kHz, 7 kHz, and 9 kHz. We will get sharper peaks in the FFT if we include more cycles of the time-domain waveform in the scope trace. This allows the scope to calculate the FFT with more data:

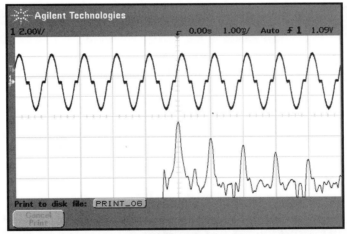

When we flip the toggle switch and close the feedback loop, the distortion magically disappears. The input and output waveforms are shown in two different scope plots:

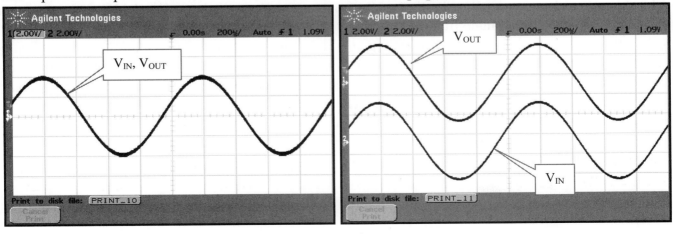

The waveforms appear to match each other almost exactly. The fast Fourier transform (FFT) of the output waveform is:

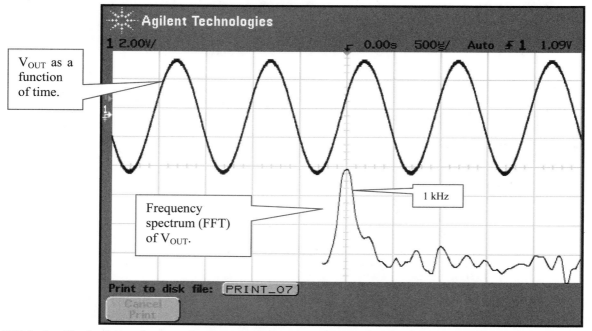

With the limited resolution shown, the output appears to contain a single frequency at 1 kHz. A higher-resolution FFT using more cycles of the output waveform is:

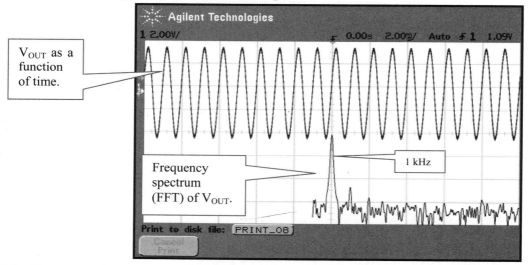

With this scope shot, we see that the FFT of the output contains a single peak at 1 kHz, indicating that the output contains only a single frequency.

We see that we can use feedback to create a very simple low-distortion amplifier. We conclude that the way to make a low-distortion amplifier is to build an amplifier with high gain, and then use feedback to reduce the gain and the distortion. The amplifier shown used ±18 V supplies so that it could supply a large-amplitude signal to a 4 Ω load. The circuit shown was able to deliver more than 50 W peak power to a 4 Ω load.

Cool Circuit VIII
Push-Pull Amplifier Thermal
Instability (Destructive)

In the previous section we discussed a push-pull amplifier and how to eliminate crossover distortion. The circuit below is the simplest form of push-pull amplifier:

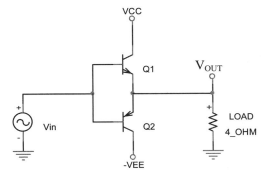

This circuit has crossover distortion because the voltage required to turn on the bipolar junction transistors must be supplied by the input source V_{IN}. When $|V_{IN}|$ is less than the turn-on voltage, neither the NPN or PNP transistors will be on, and the output will be zero for this small input. Crossover distortion is shown below:

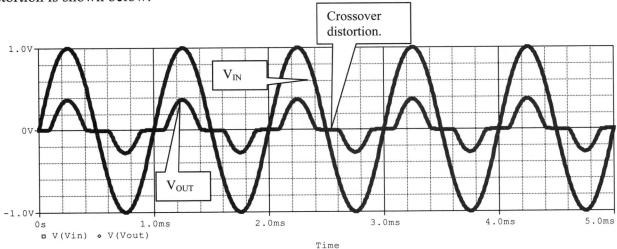

In the previous example, we eliminated this crossover distortion by using feedback. Here, we will show another method to eliminate the distortion. The problem with this method is that it is thermally unstable.

In this example, you can actually watch the devices thermally run away on a mechanical time scale. If you have a large enough power supply, the transistors may catch on fire. This makes for a cool demonstration, but the point of the demonstration is that if you are not careful, you can build potentially dangerous circuits. An audio power amplifier is meant to supply a lot of power to a speaker, and therefore has a large power supply. It is quite possible that if you build a high-power audio amplifier using a push-pull amplifier, enough power is available from the supply to burn up the power transistors. This would not be too good if this was a circuit your company designed for a vehicle or other application where a large amount of combustible fuel was present and your circuit failure could cause an explosion. But it is fun to watch the transistors run away in a controlled demonstration, and brings home the point that our circuits can fail in catastrophic ways.

VIII.A. Push-Pull Amplifier with Fixed Bias

In section VII.A.1 we saw that we could eliminate crossover distortion by providing a fixed voltage of about 0.7 V to turn on each BJT:

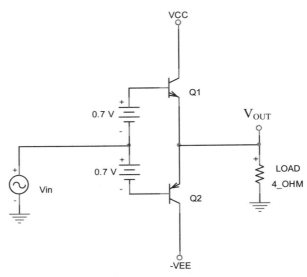

The 0.7 V sources provide the voltage necessary to bias the two BJTs so that they are always on. V_{OUT} follows V_{IN} because if we do a KVL loop from V_{IN} to V_{OUT}, the 0.7 V drop of the base-emitter junction cancels the 0.7 V increase of the fixed source. A simulation showing the input and output waveforms using TIP31 and TIP32 transistors is shown below:

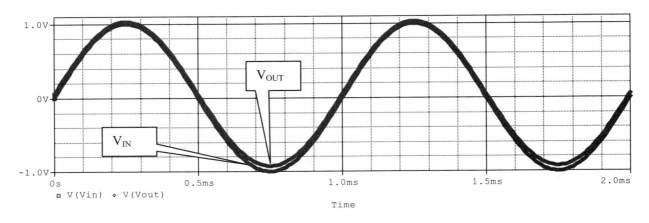

The crossover distortion is eliminated. We also note that there is a slight DC offset between the input and output waveforms because the base-emitter voltage (V_{BE}) of the NPN device is not the same as the emitter-base voltage (V_{EB}) of the PNP device. Examining the bias values of my simulation, I find that

V_{BE} is 649 mV and V_{EB} is 751 mV. Although the TIP32 is the complementary PNP device to the TIP31, the devices do not match exactly, and you should expect slight differences.

The DC sources shown are implemented in several ways. One method is to use diodes to provide the required voltage drop:

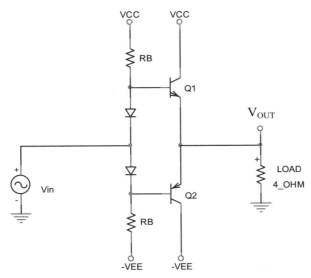

You may wonder how well the diode voltage drop matches the voltage needed by the BJTs. If the diode voltage is too small, you may see a little bit of crossover distortion. If the diode voltage drop is too high, the thermal instability and circulation current we will discuss later will be more pronounced. To closely match the bias voltage to the BJTs, we can use BJTs wired as diodes to provide the fixed bias voltage:

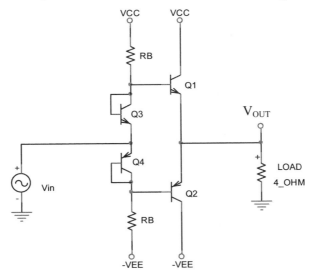

To get a close match, you might want to choose the Q_1 and Q_3 transistors to be the same model transistor, and Q_2 and Q_4 to be the same model. With similar characteristics, the base-emitter voltage drops may be closer than using diodes. The collector currents in Q_1 and Q_3, however, will be different, and since the devices are discrete, the devices will have slightly different characteristics. Thus, even when using the same devices for Q_1 and Q_3, and Q_2 and Q_4, the base-emitter voltages may not match as well as we would like; however, the arrangement does eliminate the crossover distortion. A last problem with using the same devices for Q_1 and Q_3, and for Q_2 and Q_4, is that for high-power amplifiers, Q_1 and Q_2 will be big power transistors. It seems like a waste of money to hook up a large, expensive power transistor as a diode. A big power transistor also takes up a lot of board space.

Another method of providing the bias voltage is to use a Zener to replace both diodes. It is difficult to find a Zener with the required voltage, so we will use the "Zener-like" circuit shown below:

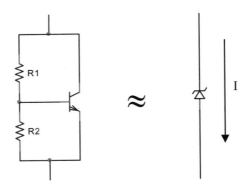

The breakdown voltage is programmable and is set by choosing R_1 and R_2. It works as a Zener only for current flowing in the direction shown. It will not behave like a diode when current flows in the direction opposite to that shown above.

We can use this circuit to bias the push-pull amplifier so that the power transistors are always on:

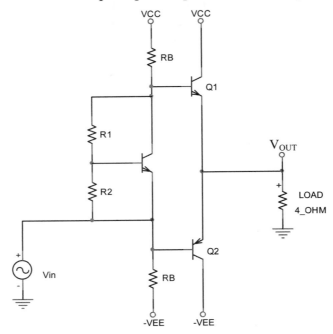

The output will have a DC offset because the input is not connected to a point that is as close to ground as it was in the previous two cases. We can eliminate this offset by using feedback as we did in section VII.B:

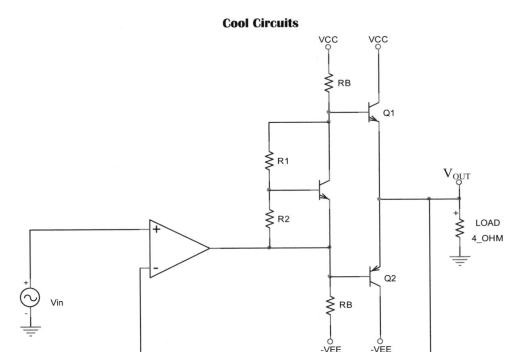

All of the methods shown will eliminate crossover distortion; however, all of them can suffer from the thermal instability and circulation current problems discussed in the next section.

VIII.B. Push-Pull Thermal Instability

There are two problems with biasing the power transistors so that they are always on. The first is that current is always flowing through the power transistors even if there is no load. Current will flow from V_{CC} to $-V_{EE}$ through the two power transistors:

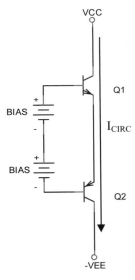

If V_{IN} is zero and Q_1 is closely matched to Q_2, the output voltage will be close to zero and no current will flow through the load. Current flows as shown and all it does is heat up Q_1 and Q_2 and waste power. Some people refer to this current as the standby current or circulation current. The standby or circulation current flows when no input is present and the output is zero.

The magnitude of the circulation current I_{CIRC} depends on how well the DC bias sources match the power transistors. If the bias sources are too large, the circulation current will be larger and waste more power. If the bias is too small, the circulation current will be less, but we may get crossover distortion.

The second problem that occurs is more subtle. Suppose that you build your push-pull amplifier using one of the bias methods shown. You power up the amplifier and measure all of the bias voltages. They all look fine and the circulation current is fairly low. You think that your design and construction are great. Suddenly, your significant other calls you on your cell phone. No, you didn't know that Jack was voted off the island. Yes, next week is our three-week anniversary. You did think that Bill and JoAnn were seeing each other, but you didn't want to mention your suspicion to anyone. What! The Beatles broke up! How could I have missed that one? Yes, you will stop by the convenience store on the way home. I did like the fragrance you were wearing the other night. It wasn't quite as nice as the one you wore when we first met. You know… The one that smelled like... What! That was someone else? Smell… something smells. Maybe tomorrow night we'll go out and watch the fire? Fire! My circuit is on fire? Smoke is coming out of the power transistors! I'll call you back….

The circuit was working fine when you checked it out. What happened? You rebuild the circuit and then measure the voltages and check the circulation current again. Everything looks fine. You leave the circuit running and then take a bathroom break. When you return, your circuit is again smoking and on the verge of flames. What is going on? You build your circuit again, test it, and verify that it works. This time you decide to sit and watch it. The bias voltages look good and the circulation current is acceptable. You sit and stare at it. Your circuit stares back. Nothing happens. You yawn. You notice that the meter that displays the circulation current has increased by a milliamp. (This is usually the current display for your main power supply.) No concern. It changed a milliamp in one minute. You sit and watch. A little while later, the current increases another milliamp. You've seen other circuits where the current fluctuates, so you are not concerned. The current increases again. This time you think that it happened a little faster than the last time. The current increases again, this time sooner than before. After a few minutes, the current is steadily increasing and it appears to be accelerating. You touch one of the power transistors (your fingers are a good thermal sensor, but the method is not recommended), and your finger gets burned. The current has now increased well over an amp. You decide to turn off the power before your circuit catches on fire again.

It turns out that the push-pull amplifier shown is thermally unstable. Even without an input, the instability will occur. The thermal instability is caused by what is referred to as the negative temperature coefficient of the bipolar junction transistor. If you recall from section I, a diode can be used as a temperature sensor because the voltage and current are dependent on temperature. For a constant current through the diode, the diode voltage changes by approximately -2 mV/°C. [1] That is, the diode voltage goes down as temperature increases. This is referred to as a negative temperature coefficient. See section I.A for a more detailed discussion on the temperature dependence of a diode.

A BJT is similar to a diode in its temperature dependence. For constant base current, as in the circuit below, the base-emitter voltage will decrease by -2 mV/°C. [15]

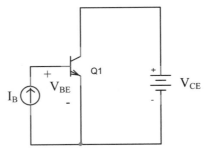

This decrease in V_{BE} due to temperature is why we say that a BJT has a negative temperature coefficient.

Also similar to a diode, if we have a constant base-emitter voltage, the collector current will double for each 10°C in device temperature. [1] If you recall, for constant diode voltage, a diode's current will double for every 10°C increase in device temperature. The circuit below maintains constant base-emitter voltage:

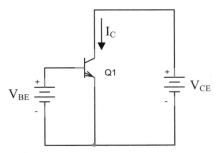

As the BJT heats up, the collector current will increase.

It turns out that the circuit above is very close to the push-pull amplifier with a constant base-emitter voltage provided to eliminate crossover distortion. Let's look at the push-pull amplifier with the input source set to zero:

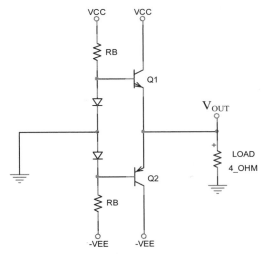

Since the output voltage follows the input voltage closely, we expect that the output voltage will be close to zero, and very little current will flow through the load. Since V_{OUT} is zero, the circuit above is equivalent to the one below for analysis purposes. We have shown two circuits: one with the load, and the equivalent circuit:

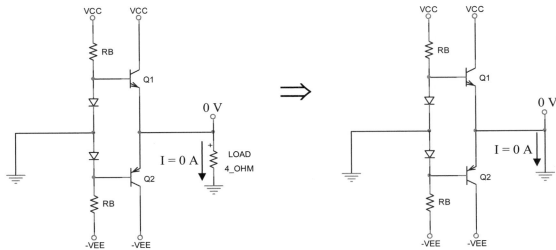

The circulating current flows from V_{CC} to $-V_{EE}$, as does the current through our bias network, I_{BIAS}:

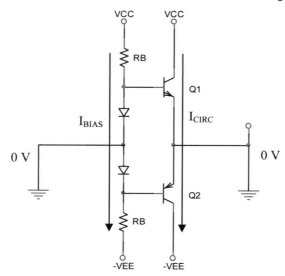

We are concerned with how the circulation current heats up the devices. Note that ground can absorb current, so for analysis purposes, the above circuit is equivalent to the one below:

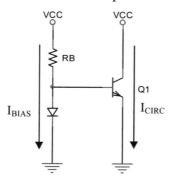

For analysis of the top half, we do not really care if the bias and circulation currents flow to ground or to the lower half of the circuit. The voltages and currents are the same for this circuit as for the actual circuit. We just need to remember that whatever happens to the top half also happens to the bottom half.

Now that we have this simplified circuit, let's see what happens. The voltage across the diode is constant and behaves like a DC voltage source. Thus, we are maintaining a constant base-emitter voltage for Q_1. When we turn on the power, some circulation current will flow. The collector-emitter voltage V_{CE} is equal to the supply voltage V_{CC}. The power dissipated by Q_1 is approximately $I_C \cdot V_{CE}$, which in this case is equal to $I_{CIRC} \cdot V_{CC}$. This power dissipation heats up Q_1. As Q_1 heats up, the collector current I_{CIRC} increases. Remember that collector current doubles for each 10 °C in device temperature. Thus, we turn on the power and the device heats up. As the device heats up, I_{CIRC} increases. As I_{CIRC} increases, the power dissipated by the device $I_{CIRC} \cdot V_{CC}$ increases. As the power dissipation increases, the device temperature increases, which in turn increases the circulation current, which in turn heats up Q_1 even more. All this time, while the BJT temperature and collector current increase, the diode voltage remains constant because we placed the diode far away from Q_1, which keeps the diode cool and the diode voltage constant. This positive feedback loop continues until the power supply for V_{CC} current limits or Q_1 burns up.

You can easily test this circuit in the lab and demonstrate its thermal instability. If you have a power supply that has a small current capability, say one amp, you will be able to observe the thermal instability, but your BJT will probably not catch on fire because you do not have enough power. If you use a 10 A power supply, you will see the current increase to a very large value and then the BJT will burn up. A picture of a toasted BJT and the corresponding test circuit are shown below:

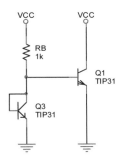

In the circuit above, Q_3 is a BJT wired as a diode to get a good match between Q_1 and Q_3. [16]

VIII.C. Fixing the Thermal Instability

There are two ways to fix the thermal instability. First, we can thermally connect the reference diode to the power transistor for which it provides the turn-on voltage. By thermally connecting the diode to the BJT, the diode heats up with the BJT. We can thermally connect the two devices by mounting them on the same heat sink and placing them physically close to one another. If you remember, a diode also has a negative temperature coefficient. As the diode heats up, its voltage goes down. In the circuit shown, Q_3 is connected as a diode. As Q_3 warms up, V_{BE} will go down.

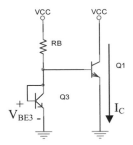

As I_C increases, the power dissipated by Q_1 warms up Q_1. Since Q_1 is thermally linked to Q_3, as Q_1 warms up, so does Q_3. As Q_3 warms up, the voltage V_{BE3} decreases. The way the circuit is wired, V_{BE3} equals V_{BE1}, so as I_C increases and heats up Q_1 and Q_3, V_{BE1} decreases. This decrease in V_{BE1} tends to reduce I_C and stabilize the circuit. By thermally connecting Q_1 and Q_3, we are adding a thermal negative feedback loop that tends to reduce I_C as Q_1 warms up. This balances the positive feedback loop that originally made the circuit thermally unstable.

A second method to stabilize the circuit requires the addition of an emitter resistor. For simplicity, we will work with the top half of the push-pull circuit; but any components we add to the top half must be added to the bottom half. We add the emitter resistor R_E as shown below:

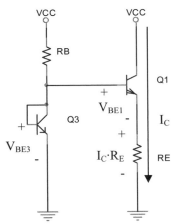

If you studied a three-resistor bias for BJTs, you may recall that this resistor adds local negative feedback and stabilizes the bias against variations in H_{FE}. In our circuit, R_E also adds negative feedback, but in this case the negative feedback stabilizes the circuit against thermal instabilities. If we assume that the emitter current is approximately equal to the collector current for Q_1, then the voltage drop across R_E is $I_C \cdot R_E$. We can then write Kirchhoff's voltage law around the base-emitter loop as:

$$V_{BE1} = V_{BE3} - I_C R_E$$

For the moment we will assume that Q_1 and Q_3 are not thermally linked, so V_{BE3} is constant. As Q_1 warms up, I_C will increase due to the thermal positive feedback discussed earlier. Because we added R_E, however, V_{BE1} will decrease as I_C increases. The decrease in V_{BE1} will tend to decrease the collector current. This again is a negative feedback loop that tends to stabilize the thermal instability.

The best method to eliminate the thermal instability is to thermally couple Q_1 and Q_3, and also add the feedback resistor R_E. Now that we have added R_E, we need to see what the entire push-pull amplifier looks like. We actually need two resistors, one for the NPN half and one for the PNP half:

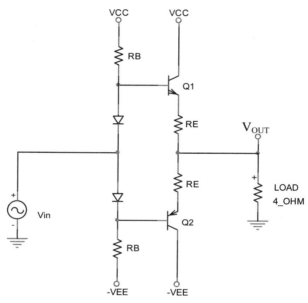

The problem with adding R_E is that it reduces the voltage gain V_{OUT}/V_{IN} and the voltage swing at the output. We should ask the question, "How large should R_E be?" If we make R_E too small, the negative feedback will be too small to balance the thermal instability and the circuit will run away thermally. As we make R_E larger, we tend to reduce the swing and voltage gain of the amplifier. You should make R_E as large as possible while still meeting the voltage swing and gain specifications required by the stage. Also, use thermal feedback to link the bias diodes to the power transistors. Just because we say "make R_E as large a possible" does not mean R_E will be a big resistor. For higher-power amplifiers, R_E can be 1 Ω or smaller.

Cool Circuit IX
Trojan Surge Protector

One day I was lying on my couch listening to my stereo and dreaming about electric race cars. How could I make mine go faster and farther without burning out the motor? Suddenly there was a loud pop, and a large amount of green smoke started spewing out of the small metal box to which my stereo was connected. Next, green smoke started spewing out of the small box into which my computer was plugged. After a few moments the smoke stopped. I opened the windows to air out the house and I heard someone yelling, "%#$%, you better see if anyone is in that house." I walked outside and a guy with a chain saw started apologizing. "That never happens," the guy waving the chain saw said, "The trees always fall right where I want them to." I lived in a small A-frame house surrounded by Ponderosa pine trees. A new neighbor had moved in next door and decided to remove a few of the trees.

"Lucky for you the power lines stopped the tree from hitting your house." I took a look at the huge tree leaning against the power wires leading up to my house.

"I guess I was lucky…" I said, "Every electronic device in my house just exploded! But hey, that tree sure looks nice hanging in those wires. Can I have the firewood?"

After removing the tree and having the power company fix the wires, I examined the circuit breakers in my house and noticed that the ones supplying the power for the boxes that started smoking were tripped to the off position. Those boxes were surge protectors that I had made to protect my electronic equipment. I removed the surge protectors and plugged in my computer and stereo. They all worked. My surge suppressors protected all of my equipment!

When I examined the surge protectors, they looked like they had caught on fire. Usually that does not happen. It turned out that when the tree hit the wires, my 115 VAC lines went up to 230 VAC. The surge protector was designed to clamp voltages above 140 VAC, so the devices clamped and started drawing a lot of current. The devices lasted long enough to trip the circuit breakers, so the power was disconnected before my electronic equipment was damaged. Since I know how to design surge protectors, I place them everywhere in my house and protect everything I can. No electronic equipment was lost, but I did lose two compact fluorescent lightbulbs. These bulbs had electronic ballasts, and because they screwed into a conventional bulb socket, I was not able to install a surge protector for them. All in all, I didn't have much damage considering what had happened.

The guy with the chain saw said that he would replace the lightbulbs for me. Now that I think of the event ten years later, I realize that he never did replace those bulbs. Good thing I never throw anything out and I still have his card.

The event that I have just described is not the usual situation for which you design a surge protector. Surges usually last a very short time, so the protector does not have to absorb a lot of power for a long time. In this example, the voltage became too high and stayed there, causing the surge protector to dissipate much more power than it was designed to handle; however, the protection devices were tough enough to handle the power until the circuit breakers turned off. Had the circuit breakers not disconnected the power, the surge protectors could have caught on fire and who knows what might have happened. We will look at designing surge suppressors for short transients, but this example shows us that our circuits may be put into situations for which they were not designed. My experience led me to change two things. First, I now place surge suppressors inside a nonflammable enclosure. I can't design them to handle the large amount of power that occurred in my falling tree incident, but I can prevent a burning circuit from catching my house on fire. Second, I moved to Indiana and now live on land that used to be a corn field. There are no trees anywhere near my house.

IX.A. Surge Suppression Devices

We will look at two devices for protecting against voltage transients. One is called a Metal Oxide Varistor, or MOV, and the other is a Transient Voltage Suppressor, or TVS. A MOV can handle higher energy spikes, but a TVS is faster than a MOV and will clamp the surge faster. [17]

A MOV looks like two back-to-back Zeners. The I-V plot of a MOV is:

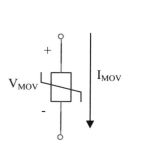
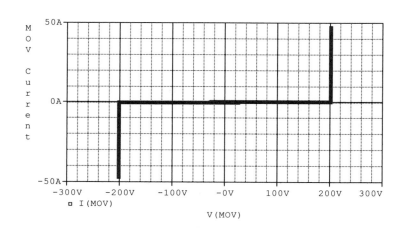

The I-V characteristic shows that a MOV behaves like two back-to-back Zeners. In this case the breakdown voltage is 200 volts:

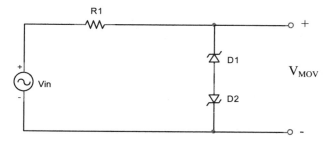

You may have used Zeners connected like this in the circuit below:

Recall how this circuit works. When $|V_{IN}|$ is less than the breakdown voltage of the Zeners, the Zeners are open circuits, and V_{OUT} equals V_{IN}. When $|V_{IN}|$ is greater than the breakdown voltage, the Zeners will clamp at the breakdown voltage, and V_{OUT} will be limited to the breakdown voltage. In other words, the magnitude of the output is limited to be less than the Zener breakdown voltage. Example waveforms for the Zener circuit are shown below. The breakdown voltage of each Zener was selected to be 200 volts:

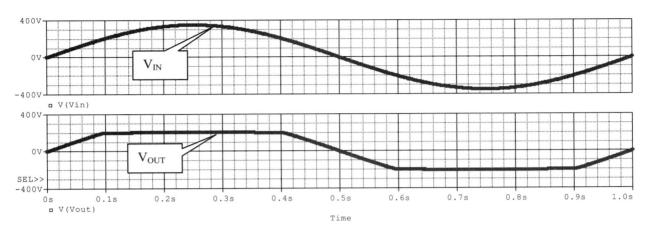

The circuit above performs the surge protection function we will be using. We will not use a Zener because Zeners are fairly slow and are not designed to handle high current spikes. Remember that when the Zener breaks down, a large current will flow through the device. Zeners are not designed to handle the extremely high currents that occur in surge protection applications. Also, Zeners are fairly slow, and they may not break down fast enough to clamp surges. Thus, we will use MOVs or TVSs for our surge suppressors. Had we used a MOV in the circuit above, the waveforms would be the same because the MOV behaves like back-to-back Zeners. A symbol for a MOV is:

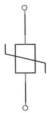

Transient Voltage Suppressors (TVS) come in unidirectional and bidirectional versions. A unidirectional TVS has the I-V characteristic of a Zener, while the bidirectional version has an I-V characteristic similar to that of the back-to-back Zeners discussed above or to that of a MOV. A unidirectional TVS uses a Zener symbol, while a bidirectional TVS has the symbol shown below:

The symbol is designed to suggest that a bidirectional TVS is similar to back-to-back Zeners. (Note that the symbols shown here are not universal; different manufacturers may use different symbols.)

A surge is a one-time event. That is, for long periods of time nothing happens. Then suddenly a single high-voltage pulse occurs. Then, nothing happens again for a long period of time. An example of a surge would be the waveform below:

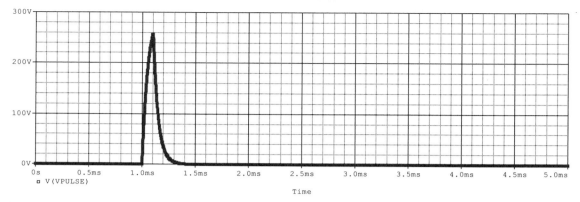

The pulse shown lasts for about 150 μs and has an amplitude of about 250 volts. Another pulse like the one shown may not occur for a very long time, possibly days or weeks. The same pulse shown on a longer time scale is:

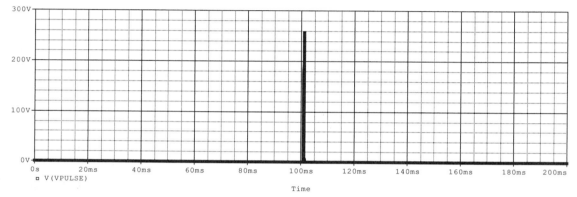

This is the type of signal a surge suppressor has to protect against. A surge can be modeled as a single pulse with a specific amplitude and pulse width; however, the pulse happens only once. How do we choose the power rating of our surge protection devices to withstand a waveform of this type?

Most devices we are familiar with have power ratings. For example, you have probably used ¼-watt resistors for your lab projects. Power ratings are for periodic signals such as the ones shown below:

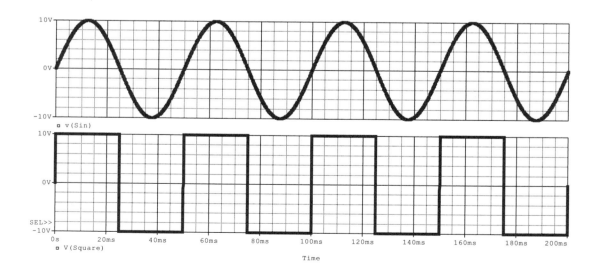

These signals are also referred to as power signals because they can continuously deliver power to a load. For example, if these were the voltages across a resistor, then we could calculate the average power delivered to the resistor as:

$$P_{AV} = \frac{1}{T} \int_0^T \frac{V(t)^2}{R} dt$$

The power absorbed by the resistor at any point in time (also called the instantaneous power) is

$$P(t) = \frac{V(t)^2}{R}$$

The instantaneous power changes with time. Sometimes the power is very high (at the peak of the sine wave), and at other times the power is zero. In a device that experiences a waveform of this type, the power continuously cycles between zero and the maximum power, and components are rated for the average power that the device absorbs. From the equation above, the average power absorbed by the resistor is the integral of the instantaneous power over one period T divided by the period.

What would happen if we tried to calculate the power carried by our surge waveform shown previously? Suppose that when the pulse occurred, the current through the surge protector was 1 amp. During the surge, the power would be the voltage times the current. The protector would absorb about 250 watts of power (250 V times 1 amp) for about 150 µs. Do we choose a 250-watt device? We also note that the power absorbed by the device is zero most of the time. The power could be zero for days or months, and then for a few microseconds the power could be huge (250 W in our example). Do we choose the surge protector to handle this peak power of 250 watts?

The surge signal shown is referred to as an energy signal. Because the signal is not periodic and the power is zero most of the time, it does not make much sense to calculate the average power. For a brief amount of time, however, the signal is dangerous and carries a large but finite amount of energy. If our pulsed waveform appeared across a resistor, the energy delivered to that resistor would be:

$$E = \int_0^\infty \frac{V(t)^2}{R} dt$$

That is, the energy is the integral over all time of the instantaneous power. Since the power is zero everywhere except during the pulse, the calculation is finite. We must choose our surge protector to handle the energy contained in the pulse, not the power at the peak of the surge. Signals may have a huge peak power, but they usually contain a finite amount of energy that we can handle.

TVSs and MOVs are rated by the amount of energy they can absorb in a single pulse, or by the amount of power they can handle for a pulse of specified width. In contrast, a typical ¼-watt resistor is rated for the amount of power it can continuously absorb. For example, a typical low-power TVS can handle a 500 W peak power for the pulse shown below:

Fig. 3 – Pulse Waveform

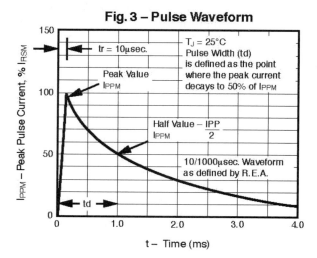

The waveform above is referred to as a 10/1000 μs waveform. This notation means that the rise time of the pulse is 10 μs and the amount of time it takes the pulse to decay to one-half of its peak value is 1000 μs. It is a standard pulse shape when talking about MOVs and TVSs. You will also see an 8x20 μs pulse shape for MOVs. This means that the rise time of the pulse is 8 μs and it takes the pulse 20 μs to decay to one-half of its peak value. The duty cycle for a typical TVS is about 0.01%. The duty cycle is the ratio of the pulse width to the period of the signal:

$$D = \frac{PW}{T} \times 100$$

In the waveform below, the duty cycle is 20 percent because the period (T) is 5 seconds and the pulse width is 1 second. Note that the waveform is periodic:

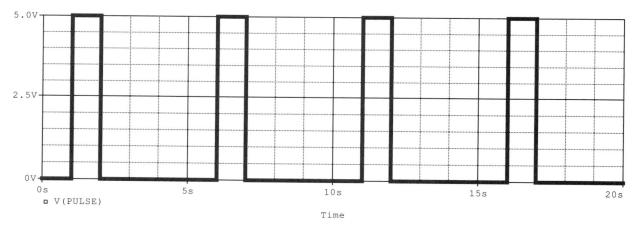

Our example TVS is rated for a duty cycle of 0.01%. If the pulse width were rated at 1 second, then at a duty cycle of 0.01%, a pulse could occur every 10,000 seconds. In our TVS, the half-value pulse width is rated at 1000 μs, so at a duty cycle of 0.01% a pulse can occur every ten seconds. We see that a TVS is a device rated for a narrow pulse that does not happen very frequently.

Typical MOVs are rated for the amount of energy they can absorb for a 10/1000 μs pulse and for a 8x20 μs pulse. The ratings are similar to TVSs, but MOVs specify an energy rating rather than a power rating. MOVs typically can handle much higher current surges than TVSs, but MOVs can only absorb a limited number of surges [18], while TVSs have no such limit.

You need to choose a TVS or a MOV to handle the largest amount of energy that you think a surge will contain, and by the breakdown voltage of the device. If we want to put a surge suppressor across the 115 VAC line, it had better not break down for normal line voltage variations, which can be as high as 127 VRMS. Note that a sine wave of 127 VRMS has a peak voltage of 180 V. MOVs specify a DC rating and an AC RMS voltage rating. For a MOV, we should choose the DC rating above 180 V and the AC rating above 127 V. A MOV with a 140 VAC rating usually satisfies both constraints. A TVS has a voltage rating where it breaks down. We need to choose a TVS that breaks down at a voltage greater than 180 V. Remember, we do not want the surge suppressor to break down and conduct for normal line voltages, which can be as high as 127 VRMS. If the line goes this high for an extended amount of time and the MOV or TVS breaks down, the MOV or TVS will continuously absorb power and will fail. Thus, we need to choose the breakdown voltage of our devices to be higher than the normal high line voltages we expect to use.

IX.B. Basic Surge Protection Circuit

If you open up an inexpensive surge-protected power strip, you will see that it is a plain power strip with a MOV across the terminals. When you plug it into an outlet, you are effectively placing the MOV in parallel with a 115 VAC source. An equivalent circuit is shown below:

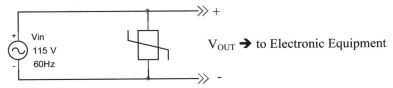

The breakdown voltage of the MOV is chosen larger than the 115 V line, so the MOV is normally an open circuit and has no effect. So, what happens when a surge appears on the voltage source and the voltage goes up to, say, 1000 V? The MOV will break down and start drawing a large amount of current. When the MOV breaks down, what will V_{OUT} do? The answer is that V_{OUT} will look like V_{IN}, and the 1000 V spike will appear on the output and damage any equipment connected to the surge suppressor. The MOV does nothing.

So why does V_{OUT} look exactly like V_{IN}? We can answer the question by looking at the Zener clipping circuits we usually study in class. An example is the circuit below:

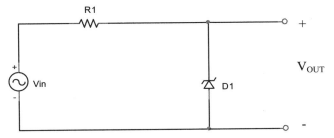

What role does R_1 play in the circuit above? One answer is that it limits the current through the Zener when it breaks down. The second is that when the Zener breaks down and limits the voltage at the output, the difference between V_{IN} and V_{OUT} appears across the resistor. Without the resistor, V_{OUT} would have to equal V_{IN} even if the Zener draws a large amount of current. Remember that in an ideal voltage source the voltage is independent of the current drawn from the source. For an ideal source, you can draw an infinite amount of current from the source and the voltage will not be affected. Thus, for the circuit below, the voltage at the output will be a sine wave no matter what the Zener does:

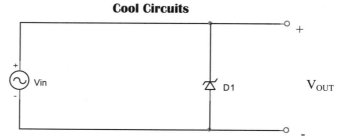

Even if the Zener breaks down and draws a lot of current, an ideal voltage source can supply that current and the voltage will be unchanged.

So now we have a mystery. How can companies sell the standard inexpensive surge protector shown below and have $25,000 equipment guarantees?

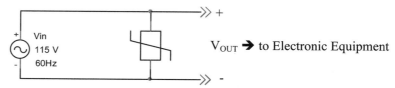

One answer is that computers and electronic equipment lose their value very quickly. Chances are it will be a long time before you are hit by a dangerous surge. Then, it will take you a long time to find a number to call to process your claim because you threw out the documentation for the surge suppressor years ago. Why keep documentation on a device that cost you only $15? Once you finally find that number and call it, you will go through twenty menu levels in an automated phone system that really does understand voice commands, just not yours. Once you find the option to talk to a human, you are put on hold and given an estimated hold time. While on hold, you will be disconnected several times, so you will have to repeat navigating through the twenty menu levels. By the time they get around to paying you, your equipment will be worthless and you will get $20 for the computer for which you once paid $2000.

The above scenario has never happened to me because I build my own surge protectors and I'm too cheap to pay myself the replacement cost of the equipment when one of them explodes. (Besides, the $2000 is already appropriated for that dinette set I've had my eye on.) Also, none of my surge protectors have ever failed. The reason given above is not the reason manufacturers can sell inexpensive surge protectors. The real reason is that they work. Why? The answer is that something in the circuit plays the role of the series resistor to drop the voltage. When a surge occurs, it usually does not originate at your wall plug. It usually originates far away and then travels down long wires to your wall outlet. An example circuit is shown below:

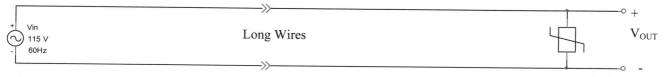

Wire has inductance. Long wires have a significant inductance, so the equivalent circuit is:

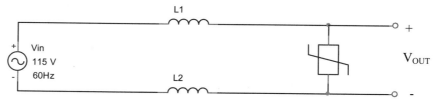

The longer the wire is, the higher the inductance will be. V_{IN} is a 60 Hz sine wave with a very narrow high-voltage pulse on top of the sine wave. Typical spikes may be on the order of a few microseconds or milliseconds. An example waveform of a surge is:

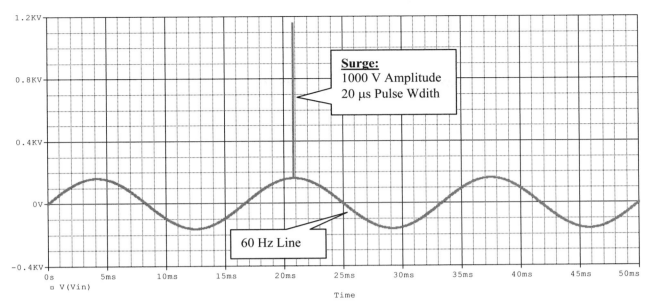

A narrow pulse contains high frequencies. Remember that the impedance of an inductor is $Z_L = j\omega L$. The inductance is small, so Z_L is negligible for the low 60 Hz frequency of the line. Thus, the line inductance does not affect the 60 Hz line that much. A surge, however, contains very high frequencies, so the impedance of the inductor to the surge is significant. Thus, the impedance of the wires to your house allows the surge protector to work. The MOV can clamp, and the difference between the 115 V source and the MOV appears across the line inductance.

Now, if lightning strikes a telephone pole far away from your house, there is a lot of wire between your outlets and the source of the surge. The line inductance is significant and the surge suppressor can be effective because of the large line inductance. If, however, lightning directly strikes your house, there is very little wire between the source and your outlets; so the line inductance is reduced because of the short distance the surge has to travel. In this case, the surge suppressor will be less effective. And, if lightning directly strikes your computer, chances are the surge suppressor will completely fail. With any luck, you will not be using the computer when this happens.

So, a MOV directly across your outlets can be a very effective surge suppressor because of the line inductance. A typical MOV costs less than a dollar, and you can place these devices across all of your outlets.

IX.C. Filtered Surge Suppressor Circuit

The next level up from the basic surge suppressor is the filtered surge suppressor. This circuit is basically a capacitor in parallel with a MOV. I use a 0.1 µF capacitor because a larger capacitor is expensive and you can hear it buzz at 60 Hz:

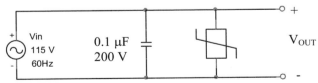

The capacitor should be a ceramic or polyester capacitor with a voltage rating of 200 V or higher. It must also be a non-polarized capacitor because the voltage across it will change polarity.

Once again, the line inductance allows this surge suppressor to work. Without a series impedance to drop the voltage, V_{OUT} will equal V_{IN} no matter what the capacitor and MOV do. Thus, we will draw the circuit with the line inductance:

L1

+   Vin
—●   115 V   0.1 µF   V~OUT~
—   60Hz   200 V

L2

We must remember that the impedance of a capacitor is $Z_C = 1/j\omega C$. The 60 Hz line frequency is so low that the capacitor can be considered an open circuit to 60 Hz. To 60 Hz the inductors are short circuits and the capacitor is an open circuit, so they have no effect at the 60 Hz line frequency. Thus, at 60 Hz, V_{OUT} equals V_{IN}.

A surge is high frequency, and the impedance of the capacitor is small at high frequency. As we learned in the last section, the impedance of the inductors is large at high frequency. We have a reactive voltage divider made up of the inductors and the capacitor. The inductors have a high impedance and the capacitor has a low impedance. Our output is across the low-impedance capacitor. So the circuit acts like a voltage divider and reduces the size of the voltage spike at the output. Thus, even without the MOV, the capacitor would reduce the size of the surge at the output. If, after this filtering, the output voltage is too high, the MOV will break down and clamp the output.

Another benefit of the capacitor is that it slows the rise time of the surge. A MOV or a TVS takes time to break down. Slowing down the rise time of a surge will give the MOV or TVS more time to clamp the surge.

A last way to look at the line inductance and capacitor is that it is a low-pass filter. The filter passes frequencies at 60 Hz, but reduces signals at high frequency. The low-pass filter tends to filter out or reduce the size of the high-frequency surge. Anything that makes it through the filter is clamped by the MOV.

This is the surge suppressor I place on most of my outlets. I place a 140 VAC MOV in parallel with a 0.1 µF capacitor in all important outlets. You can easily place these devices behind the outlet and then place them in a metal wall outlet box. You need to place the circuit in a fireproof enclosure for the rare event that the voltage goes above the MOV breakdown voltage and stays there. A picture of a MOV and capacitor mounted on an outlet is shown below. **Note that the circuit is placed in a fireproof metal outlet box in case a tree hits your house and the MOV catches on fire.**

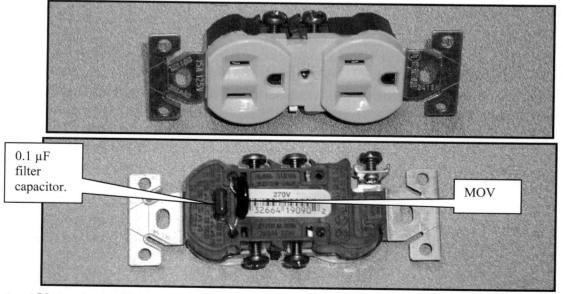

Important Note: According to UL1449, a MOV that is protecting a 110 – 120 V line should be able to handle a constant line voltage of 240 V AC RMS. [19] A common occurrence is the failure of a

transformer center tap, which will result in a 120 V line jumping up to a continuous voltage of 240 V. A MOV can catch on fire when this fault occurs, so you need either to choose a MOV that does not break down at 240 V AC RMS, or to install thermal protection to prevent the MOV from catching on fire. We did not consider this problem in our designs, as our MOVs were chosen to break down at 140 V. Note that a 240 V MOV may not protect some of the equipment we wish to protect.

IX.D. The Trojan Surge Protector

So, how can we make the simple surge suppressors of the previous two sections better? We realize that it is the line inductance that makes the surge suppressor work. To make the suppressor more effective we will add series inductors to help the line inductance. We will use the circuit below:

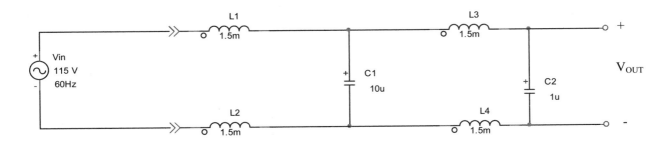

I chose 1.5 mH inductors because I happened to have 1.5 mH inductors rated for 20 amps lying around my barn, and I was not using them. We have added a serious amount of inductance, so this circuit should be much more effective. The circuit is actually a fourth order low-pass filter. The cutoff frequency was chosen to be a little bit higher than 60 Hz so that the filter passes 60 Hz, but severely attenuates the higher frequencies contained in a surge. A frequency plot of the gain V_{OUT}/V_{IN} with a 10 Ω load is shown below. A 10 Ω load will draw about 11.5 amps, which is close to the full power rating of a circuit breaker in your home.

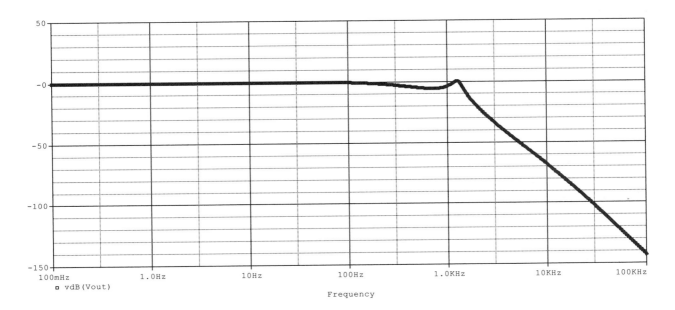

At 60 Hz the gain is 0 dB, or a gain of 1. Thus, the filter passes the 60 Hz line frequency almost unaffected. The filter starts rolling off at frequencies above 100 Hz. At high frequencies, the slope of the filter is close to –80 dB per decade. Thus, the filter kills high-frequency signals.

The next question is, what does this filter do to the 1000 V, 20 μs pulse? We will run a simulation and see what the output looks like. The results of the simulation are shown below:

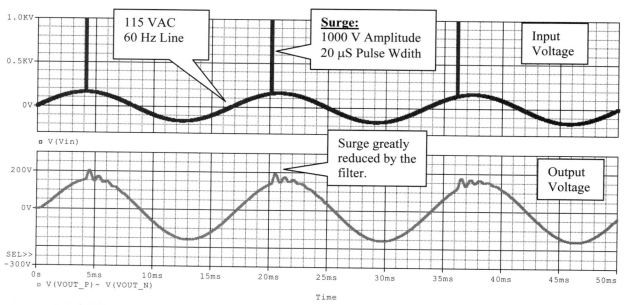

We see that the filter, by itself, greatly reduces the spike.

Next, we add MOVs and TVSs in parallel with the capacitors to clamp any spikes that make it through the filter. The full Trojan surge suppressor is shown below:

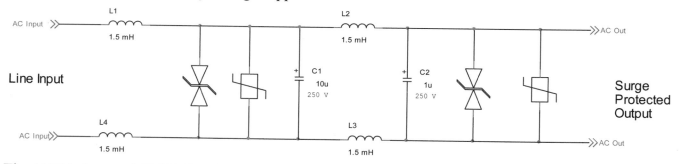

The MOVs have a 140 VAC rating and the TVSs have a minimum voltage rating of at least 180 V so that they can handle a 20% high line voltage without clamping. We are using both MOVs and TVSs to take advantage of the benefits of both devices. TVSs react faster than MOVs, so they can clamp a surge faster. [17] MOVs have higher current ratings than TVSs, so they can absorb the higher-energy surges.

Pictures of my Trojan surge suppressor are shown below:

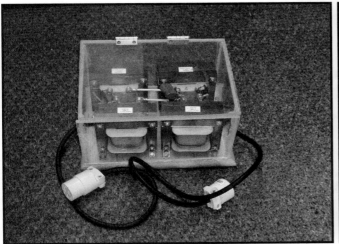 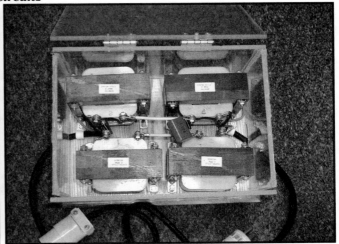

Note that the inductors are huge. The inductors are 1.5 mH inductors designed to handle a 20-amp rms current. They cost about $50 each and I had them custom made for another project. They were lying around my barn not being used, so I used them in the Trojan surge suppressor. They make the suppressor very expensive, but hey, I wasn't using them. You can use smaller inductors and still make very effective surge suppressors. I recommend smaller inductors because you can get very inexpensive inductors in the microhenry range. The filter will not be as good as the one shown, but the surge suppressor will be much more effective than the basic and filtered surge suppressors shown earlier.

I use the Trojan suppressor for my sensitive equipment. One day, we were having a party and lightning struck my house. It didn't strike several hundred feet away; it actually struck my house and caught a tree on fire. (You know, with all of these lightning problems, circuits catching on fire, and high-voltage shocking circuits, it's a wonder I'm still alive.) All of my equipment connected to the Trojan survived. The only loss was my stereo receiver, which was not powered through the Trojan, but was plugged directly into a wall outlet. Also, it was not turned on. (We have great parties because we never turn on the stereo.) So, why did my stereo get fried even though it was not turned on? The answer is that it is on all the time, but in a low-power mode. For the remote control to work when a device is "off," there must be a small power supply in the device running continuously so that it can receive signals from the remote control. Even though the main power supply for the stereo was off, a small standby power supply was running to power the electronics needed to receive information from the remote control. It was this portion of the stereo that was fried. Oh well… That's what I get for not using a surge suppressor on all of my electronic equipment.

Important Note: A filtered surge suppressor, such as the one shown here, can have a resonant frequency where the output voltage is larger than the input voltage. If you use the filtered surge suppressor to power a high-frequency switching circuit, such as the electronic hot dog cooker, you may inadvertently hit the resonant frequency and cause bad things to happen, such as an output voltage higher than 115 VAC. More simulation of the Trojan is needed to see whether a resonant condition can occur, and what problems it may cause.

Cool Circuits
Appendix A
Datasheets

New Product

SMA5J5.0 thru 40CA

Vishay Semiconductors
formerly General Semiconductor

High Power Density Surface Mount TRANSZORB®
Transient Voltage Suppressors

Stand-off Voltage 5.0 to 40 V
Peak Pulse Power 500 W

DO-214AC (SMA)

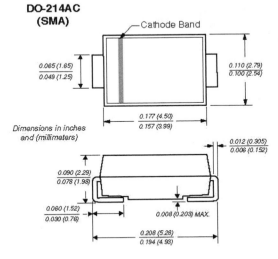

Dimensions in inches and (millimeters)

Mounting Pad Layout

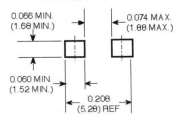

Features

- Plastic package has Underwriters Laboratory Flammability Classification 94V-0
- Ideal for ESD protection of data lines in accordance with IEC 1000-4-2 (IEC801-2)
- Ideal for EFT protection of data lines in accordance with IEC 1000-4-4 (IEC801-4)
- Low profile package with built-in strain relief for surface mounted applications
- Glass passivated junction
- Low incremental surge resistance, excellent clamping capability
- 500W peak pulse power capability with a 10/1000µs waveform, repetition rate (duty cycle): 0.01%
- Very fast response time

Mechanical Data

Case: JEDEC DO-214AC molded plastic over passivated chip
Terminals: Solder plated, solderable per MIL-STD-750, Method 2026
High temperature soldering guaranteed: 250°C/10 seconds at terminals
Polarity: For uni-directional types the band denotes the cathode, which is positive with respect to the anode under normal TVS operation
Mounting Position: Any   **Weight:** 0.002oz., 0.064g

Devices for Bidirectional Applications

For bi-directional devices, use suffix C or CA (e.g. SMA5J10CA). Electrical characteristics apply in both directions.

Maximum Ratings & Thermal Characteristics
Ratings at 25°C ambient temperature unless otherwise specified.

Parameter	Symbol	Value	Unit
Peak pulse power dissipation with a 10/1000µs waveform[1,2] (see fig. 1)	P_{PPM}	500	W
Peak pulse current with a 10/1000µs waveform[1]	I_{PPM}	See Next Table	A
Peak forward surge current 8.3ms single half sine-wave uni-directional only[2]	I_{FSM}	40	A
Typical thermal resistance, junction to ambient[3]	$R_{\theta JA}$	80	°C/W
Typical thermal resistance, junction to lead	$R_{\theta JL}$	25	°C/W
Operating junction and storage temperature range	T_J, T_{STG}	−55 to +150	°C

Notes: (1) Non-repetitive current pulse, per Fig. 3 and derated above $T_A = 25°C$ per Fig. 2
(2) Mounted on 0.2 x 0.2" (5.0 x 5.0mm) copper pads to each terminal
(3) Mounted on minimum recommended pad layout

SMA5J5.0 thru 40CA

Vishay Semiconductors
formerly General Semiconductor

Electrical Characteristics
Ratings at 25°C ambient temperature unless otherwise specified. $V_F = 3.5V$ at $I_F = 25A$ (uni-directional only)

Device Type	Device Marking Code		Breakdown Voltage $V_{(BR)}$ (V)[1]		Test Current at I_T (mA)	Stand-off Voltage V_{WM} (V)	Maximum Reverse Leakage at V_{WM} I_D (μA)[3]	Maximum Peak Pulse Surge Current I_{PPM} (A)[2]	Maximum Clamping Voltage at I_{PPM} V_C (V)
	UNI	BI	Min	Max					
SMA5J5.0	5AD	5AD	6.40	7.82	10	5.0	800	52.1	9.6
SMA5J5.0A[5]	5AE	5AE	6.40	7.07	10	5.0	800	54.3	9.2
SMA5J6.0	5AF	5AF	6.67	8.15	10	6.0	800	43.9	11.4
SMA5J6.0A	5AG	5AG	6.67	7.37	10	6.0	800	48.5	10.3
SMA5J6.5	5AH	5AH	7.22	8.82	10	6.5	500	40.7	12.3
SMA5J6.5A	5AK	5AK	7.22	7.98	10	6.5	500	44.6	11.2
SMA5J7.0	5AL	5AL	7.78	9.51	10	7.0	200	37.6	13.3
SMA5J7.0A	5AM	5AM	7.78	8.60	10	7.0	200	41.7	12.0
SMA5J7.5	5AN	5AN	8.33	10.2	1.0	7.5	100	35.0	14.3
SMA5J7.5A	5AP	5AP	8.33	9.21	1.0	7.5	100	38.8	12.9
SMA5J8.0	5AQ	5AQ	8.89	10.9	1.0	8.0	50	33.3	15.0
SMA5J8.0A	5AR	5AR	8.89	9.83	1.0	8.0	50	36.8	13.6
SMA5J8.5	5AS	5AS	9.44	11.5	1.0	8.5	10	31.4	15.9
SMA5J8.5A	5AT	5AT	9.44	10.4	1.0	8.5	10	34.7	14.4
SMA5J9.0	5AU	5AU	10.0	12.2	1.0	9.0	5.0	29.6	16.9
SMA5J9.0A	5AV	5AV	10.0	11.1	1.0	9.0	5.0	32.5	15.4
SMA5J10	5AW	5AW	11.1	13.6	1.0	10	1.0	26.6	18.8
SMA5J10A	5AX	5AX	11.1	12.3	1.0	10	1.0	29.4	17.0
SMA5J11	5AY	5AY	12.2	14.9	1.0	11	1.0	24.9	20.1
SMA5J11A	5AZ	5AZ	12.2	13.5	1.0	11	1.0	27.5	18.2
SMA5J12	5BD	5BD	13.3	16.3	1.0	12	1.0	22.7	22.0
SMA5J12A	5BE	5BE	13.3	14.7	1.0	12	1.0	25.1	19.9
SMA5J13	5BF	5BF	14.4	17.6	1.0	13	1.0	21.0	23.8
SMA5J13A	5BG	5BG	14.4	15.9	1.0	13	1.0	23.3	21.5
SMA5J14	5BH	5BH	15.6	19.1	1.0	14	1.0	19.4	25.8
SMA5J14A	5BK	5BK	15.6	17.2	1.0	14	1.0	21.6	23.2
SMA5J15	5BL	5BL	16.7	20.4	1.0	15	1.0	18.6	26.9
SMA5J15A	5BM	5BM	16.7	18.5	1.0	15	1.0	20.5	24.4
SMA5J16	6BN	5BN	17.8	21.8	1.0	16	1.0	17.4	28.8
SMA5J16A	5BP	5BP	17.8	19.7	1.0	16	1.0	19.2	26.0
SMA5J17	5BQ	5BQ	18.9	23.1	1.0	17	1.0	16.4	30.5
SMA5J17A	5BR	5BR	18.9	20.9	1.0	17	1.0	18.1	27.6
SMA5J18	5BS	5BS	20.0	24.4	1.0	18	1.0	15.5	32.2
SMA5J18A	5BT	5BT	20.0	22.1	1.0	18	1.0	17.1	29.2
SMA5J20	5BU	5BU	22.2	27.1	1.0	20	1.0	14.0	35.8
SMA5J20A	5BV	5BV	22.2	24.5	1.0	20	1.0	15.4	32.4
SMA5J22	5BW	5BW	24.4	29.8	1.0	22	1.0	12.7	39.4
SMA5J22A	5BX	5BX	24.4	26.9	1.0	22	1.0	14.1	35.5
SMA5J24	5BY	5BY	26.7	32.6	1.0	24	1.0	11.6	43.0
SMA5J24A	5BZ	5BZ	26.7	29.5	1.0	24	1.0	12.9	38.9
SMA5J26	5CD	5CD	28.9	35.3	1.0	26	1.0	10.7	46.6
SMA5J26A	5CE	5CE	28.9	31.9	1.0	26	1.0	11.9	42.1
SMA5J28	5CF	5CF	31.1	38.0	1.0	28	1.0	10.0	50.0
SMA5J28A	5CG	5CG	31.1	34.4	1.0	28	1.0	11.0	45.4
SMA5J30	5CH	5CH	33.3	40.7	1.0	30	1.0	9.3	53.5
SMA5J30A	5CK	5CK	33.3	36.8	1.0	30	1.0	10.3	48.4

Notes: (1) $V_{(BR)}$ measured after I_T applied for 300μs square wave pulse or equivalent
(2) Surge current waveform per Fig. 3 and derate per Fig. 2
(3) For bi-directional types having V_{WM} of 10 Volts and less, the I_D limit is doubled
(4) All terms and symbols are consistent with ANSI/IEEE C62.35
(5) For the bidirectional SMA5J5.0CA, the maximum $V_{(BR)}$ is 7.25V.

Document Number 88875
11-Mar-04

SMA5J5.0 thru 40CA

Vishay Semiconductors
formerly General Semiconductor

Ratings and
Characteristic Curves (T_A = 25°C unless otherwise noted)

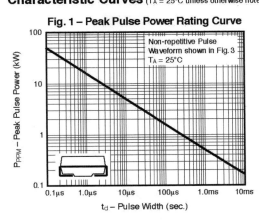

Fig. 1 – Peak Pulse Power Rating Curve

Non-repetitive Pulse Waveform shown in Fig. 3
T_A = 25°C

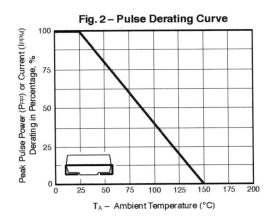

Fig. 2 – Pulse Derating Curve

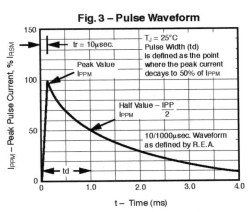

Fig. 3 – Pulse Waveform

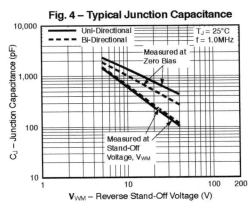

Fig. 4 – Typical Junction Capacitance

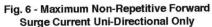

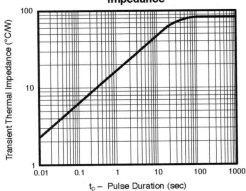

Fig. 5 – Typical Transient Thermal Impedance

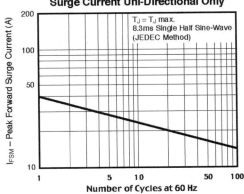

Fig. 6 - Maximum Non-Repetitive Forward Surge Current Uni-Directional Only

1N6267A Series

1500 Watt Mosorb™ Zener Transient Voltage Suppressors

Unidirectional*

Mosorb devices are designed to protect voltage sensitive components from high voltage, high–energy transients. They have excellent clamping capability, high surge capability, low zener impedance and fast response time. These devices are ON Semiconductor's exclusive, cost-effective, highly reliable Surmetic™ axial leaded package and are ideally-suited for use in communication systems, numerical controls, process controls, medical equipment, business machines, power supplies and many other industrial/consumer applications, to protect CMOS, MOS and Bipolar integrated circuits.

Specification Features:
- Working Peak Reverse Voltage Range – 5.8 V to 214 V
- Peak Power – 1500 Watts @ 1 ms
- ESD Rating of Class 3 (>16 KV) per Human Body Model
- Maximum Clamp Voltage @ Peak Pulse Current
- Low Leakage < 5 μA Above 10 V
- UL 497B for Isolated Loop Circuit Protection
- Response Time is Typically < 1 ns

Mechanical Characteristics:
CASE: Void-free, transfer-molded, thermosetting plastic
FINISH: All external surfaces are corrosion resistant and leads are readily solderable
MAXIMUM LEAD TEMPERATURE FOR SOLDERING PURPOSES: 230°C, 1/16″ from the case for 10 seconds
POLARITY: Cathode indicated by polarity band
MOUNTING POSITION: Any

ON Semiconductor®

http://onsemi.com

Cathode ——▷|—— Anode

AXIAL LEAD
CASE 41A
PLASTIC

L
1N6
xxxA
1.5KE
xxxA
YYWW

L = Assembly Location
1N6xxxA = JEDEC Device Code
1.5KExxxA = ON Device Code
YY = Year
WW = Work Week

MAXIMUM RATINGS

Rating	Symbol	Value	Unit
Peak Power Dissipation (Note 1) @ $T_L \leq 25°C$	P_{PK}	1500	Watts
Steady State Power Dissipation @ $T_L \leq 75°C$, Lead Length = 3/8″ Derated above $T_L = 75°C$	P_D	5.0 / 20	Watts / mW/°C
Thermal Resistance, Junction–to–Lead	$R_{\theta JL}$	20	°C/W
Forward Surge Current (Note 2) @ $T_A = 25°C$	I_{FSM}	200	Amps
Operating and Storage Temperature Range	T_J, T_{stg}	– 65 to +175	°C

1. Nonrepetitive current pulse per Figure 5 and derated above $T_A = 25°C$ per Figure 2.
2. 1/2 sine wave (or equivalent square wave), PW = 8.3 ms, duty cycle = 4 pulses per minute maximum.
*Please see 1.5KE6.8CA to 1.5KE250CA for Bidirectional Devices

ORDERING INFORMATION

Device	Package	Shipping
1.5KExxxA	Axial Lead	500 Units/Box
1.5KExxxARL4	Axial Lead	1500/Tape & Reel
1N6xxxA	Axial Lead	500 Units/Box
1N6xxxARL4*	Axial Lead	1500/Tape & Reel

*1N6302A Not Available in 1500/Tape & Reel

Devices listed in **_bold, italic_** are ON Semiconductor **Preferred** devices. **Preferred** devices are recommended choices for future use and best overall value.

© Semiconductor Components Industries, LLC, 2002
June, 2002 – Rev. 5

Publication Order Number:
1N6267A/D

1N6267A Series

ELECTRICAL CHARACTERISTICS (T_A = 25°C unless otherwise noted, V_F = 3.5 V Max., I_F (Note 3) = 100 A)

Symbol	Parameter
I_{PP}	Maximum Reverse Peak Pulse Current
V_C	Clamping Voltage @ I_{PP}
V_{RWM}	Working Peak Reverse Voltage
I_R	Maximum Reverse Leakage Current @ V_{RWM}
V_{BR}	Breakdown Voltage @ I_T
I_T	Test Current
ΘV_{BR}	Maximum Temperature Coefficient of V_{BR}
I_F	Forward Current
V_F	Forward Voltage @ I_F

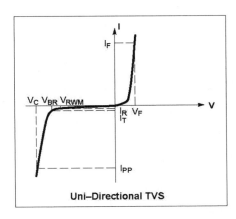

Uni–Directional TVS

1N6267A Series

ELECTRICAL CHARACTERISTICS (T_A = 25°C unless otherwise noted, V_F = 3.5 V Max. @ I_F (Note 3) = 100 A)

Device	JEDEC Device (Note 4)	V_{RWM} (Note 5) (Volts)	I_R @ V_{RWM} (μA)	Breakdown Voltage				V_C @ I_{PP} (Note 7)		ΘV_{BR}
				V_{BR} (Note 6) (Volts)			@ I_T	V_C	I_{PP}	
				Min	Nom	Max	(mA)	(Volts)	(A)	(%/°C)
1.5KE6.8A	*1N6267A*	*5.8*	*1000*	*6.45*	*6.8*	*7.14*	*10*	*10.5*	*143*	*0.057*
1.5KE7.5A	1N6268A	6.4	500	7.13	7.5	7.88	10	11.3	132	0.061
1.5KE8.2A	1N6269A	7.02	200	7.79	8.2	8.61	10	12.1	124	0.065
1.5KE9.1A	1N6270A	7.78	50	8.65	9.1	9.55	1	13.4	112	0.068
1.5KE10A	1N6271A	8.55	10	9.5	10	10.5	1	14.5	103	0.073
1.5KE11A	1N6272A	9.4	5	10.5	11	11.6	1	15.6	96	0.075
1.5KE12A	1N6273A	10.2	5	11.4	12	12.6	1	16.7	90	0.078
1.5KE13A	1N6274A	11.1	5	12.4	13	13.7	1	18.2	82	0.081
1.5KE15A	*1N6275A*	*12.8*	*5*	*14.3*	*15*	*15.8*	*1*	*21.2*	*71*	*0.084*
1.5KE16A	1N6276A	13.6	5	15.2	16	16.8	1	22.5	67	0.086
1.5KE18A	1N6277A	15.3	5	17.1	18	18.9	1	25.2	59.5	0.088
1.5KE20A	1N6278A	17.1	5	19	20	21	1	27.7	54	0.09
1.5KE22A	*1N6279A*	*18.8*	*5*	*20.9*	*22*	*23.1*	*1*	*30.6*	*49*	*0.092*
1.5KE24A	*1N6280A*	*20.5*	*5*	*22.8*	*24*	*25.2*	*1*	*33.2*	*45*	*0.094*
1.5KE27A	*1N6281A*	*23.1*	*5*	*25.7*	*27*	*28.4*	*1*	*37.5*	*40*	*0.096*
1.5KE30A	*1N6282A*	*25.6*	*5*	*28.5*	*30*	*31.5*	*1*	*41.4*	*36*	*0.097*
1.5KE33A	*1N6283A*	*28.2*	*5*	*31.4*	*33*	*34.7*	*1*	*45.7*	*33*	*0.098*
1.5KE36A	1N6284A	30.8	5	34.2	36	37.8	1	49.9	30	0.099
1.5KE39A	*1N6285A*	*33.3*	*5*	*37.1*	*39*	*41*	*1*	*53.9*	*28*	*0.1*
1.5KE43A	1N6286A	36.8	5	40.9	43	45.2	1	59.3	25.3	0.101
1.5KE47A	1N6287A	40.2	5	44.7	47	49.4	1	64.8	23.2	0.101
1.5KE51A	*1N6288A*	*43.6*	*5*	*48.5*	*51*	*53.6*	*1*	*70.1*	*21.4*	*0.102*
1.5KE56A	1N6289	*47.8*	*5*	*53.2*	*56*	*58.8*	*1*	*77*	*19.5*	*0.103*
1.5KE62A	1N6290A	53	5	58.9	62	65.1	1	85	17.7	0.104
1.5KE68A	1N6291A	58.1	5	64.6	68	71.4	1	92	16.3	0.104
1.5KE75A	1N6292A	64.1	5	71.3	75	78.8	1	103	14.6	0.105
1.5KE82A	1N6293A	70.1	5	77.9	82	86.1	1	113	13.3	0.105
1.5KE91A	1N6294A	77.8	5	86.5	91	95.5	1	125	12	0.106
1.5KE100A	1N6295A	85.5	5	95	100	105	1	137	11	0.106
1.5KE110A	1N6296A	94	5	105	110	116	1	152	9.9	0.107
1.5KE120A	1N6297A	102	5	114	120	126	1	165	9.1	0.107
1.5KE130A	1N6298A	111	5	124	130	137	1	179	8.4	0.107
1.5KE150A	1N6299A	128	5	143	150	158	1	207	7.2	0.108
1.5KE160A	1N6300A	136	5	152	160	168	1	219	6.8	0.108
1.5KE170A	1N6301A	145	5	162	170	179	1	234	6.4	0.108
1.5KE180A	1N6302A*	154	5	171	180	189	1	246	6.1	0.108
1.5KE200A	1N6303A	171	5	190	200	210	1	274	5.5	0.108
1.5KE220A		185	5	209	220	231	1	328	4.6	0.109
1.5KE250A		214	5	237	250	263	1	344	5	0.109

3. 1/2 sine wave (or equivalent square wave), PW = 8.3 ms, duty cycle = 4 pulses per minute maximum.
4. Indicates JEDEC registered data
5. A transient suppressor is normally selected according to the maximum working peak reverse voltage (V_{RWM}), which should be equal to or greater than the dc or continuous peak operating voltage level.
6. V_{BR} measured at pulse test current I_T at an ambient temperature of 25°C
7. Surge current waveform per Figure 5 and derate per Figures 1 and 2.
*Not Available in the 1500/Tape & Reel

1N6267A Series

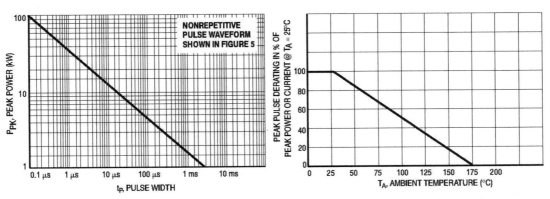

Figure 1. Pulse Rating Curve

Figure 2. Pulse Derating Curve

1N6373, ICTE-5, MPTE-5,
through
1N6389, ICTE-45, C, MPTE-45, C

1N6267A/1.5KE6.8A
through
1N6303A/1.5KE200A

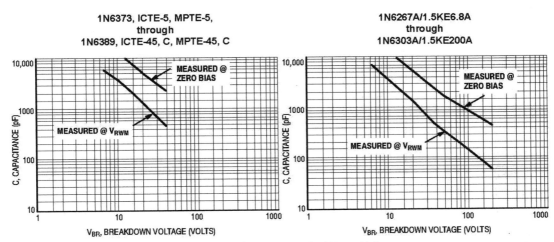

Figure 3. Capacitance versus Breakdown Voltage

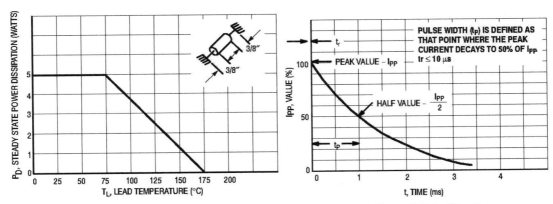

Figure 4. Steady State Power Derating

Figure 5. Pulse Waveform

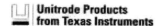
**Unitrode Products
from Texas Instruments**

UC1525A/27A
UC2525A/27A
UC3525A/27A

Regulating Pulse Width Modulators

FEATURES

- 8 to 35V Operation

- 5.1V Reference Trimmed to ±1%

- 100Hz to 500kHz Oscillator Range

- Separate Oscillator Sync Terminal

- Adjustable Deadtime Control

- Internal Soft-Start

- Pulse-by-Pulse Shutdown

- Input Undervoltage Lockout with Hysteresis

- Latching PWM to Prevent Multiple Pulses

- Dual Source/Sink Output Drivers

DESCRIPTION

The UC1525A/1527A series of pulse width modulator integrated circuits are designed to offer improved performance and lowered external parts count when used in designing all types of switching power supplies. The on-chip +5.1V reference is trimmed to ±1% and the input common-mode range of the error amplifier includes the reference voltage, eliminating external resistors. A sync input to the oscillator allows multiple units to be slaved or a single unit to be synchronized to an external system clock. A single resistor between the C_T and the discharge terminals provides a wide range of dead-time adjustment. These devices also feature built-in soft-start circuitry with only an external timing capacitor required. A shutdown terminal controls both the soft-start circuitry and the output stages, providing instantaneous turn off through the PWM latch with pulsed shutdown, as well as soft-start recycle with longer shutdown commands. These functions are also controlled by an undervoltage lockout which keeps the outputs off and the soft-start capacitor discharged for sub-normal input voltages. This lockout circuitry includes approximately 500mV of hysteresis for jitter-free operation. Another feature of these PWM circuits is a latch following the comparator. Once a PWM pulse has been terminated for any reason, the outputs will remain off for the duration of the period. The latch is reset with each clock pulse. The output stages are totem-pole designs capable of sourcing or sinking in excess of 200mA. The UC1525A output stage features NOR logic, giving a LOW output for an OFF state. The UC1527A utilizes OR logic which results in a HIGH output level when OFF.

BLOCK DIAGRAM

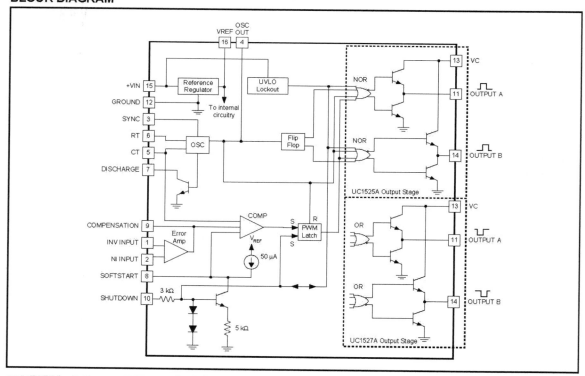

SLUS191A - February 1997 - Revised April 2004

TIP31, TIP31A, TIP31B, TIP31C, (NPN), TIP32, TIP32A, TIP32B, TIP32C, (PNP)

Complementary Silicon Plastic Power Transistors

Designed for use in general purpose amplifier and switching applications.

- Collector-Emitter Saturation Voltage -
  $V_{CE(sat)} = 1.2$ Vdc (Max) @ $I_C = 3.0$ Adc

- Collector-Emitter Sustaining Voltage -
  $V_{CEO(sus)} = 40$ Vdc (Min) - TIP31, TIP32
  $= 60$ Vdc (Min) - TIP31A, TIP32A
  $= 80$ Vdc (Min) - TIP31B, TIP32B
  $= 100$ Vdc (Min) - TIP31C, TIP32C

- High Current Gain - Bandwidth Product
  $f_T = 3.0$ MHz (Min) @ $I_C = 500$ mAdc

- Compact TO-220 AB Package

ON Semiconductor®

http://onsemi.com

**3 AMPERE
POWER TRANSISTORS
COMPLEMENTARY
SILICON
40-60-80-100 VOLTS
40 WATTS**

MAXIMUM RATINGS

Rating		Symbol	Value	Unit
Collector-Emitter Voltage TIP31, TIP32 TIP31A, TIP32A TIP31B, TIP32B TIP31C, TIP32C		V_{CEO}	40 60 80 100	Vdc
Collector-Base Voltage TIP31, TIP32 TIP31A, TIP32A TIP31B, TIP32B TIP31C, TIP32C		V_{CB}	40 60 80 100	Vdc
Emitter-Base Voltage		V_{EB}	5.0	Vdc
Collector Current	Continuous Peak	I_C	3.0 5.0	Adc
Base Current		I_B	1.0	Adc
Total Power Dissipation @ $T_C = 25°C$ Derate above 25°C		P_D	40 0.32	Watts W/°C
Total Power Dissipation @ $T_A = 25°C$ Derate above 25°C		P_D	2.0 0.016	Watts W/°C
Unclamped Inductive Load Energy (Note 1)		E	32	mJ
Operating and Storage Junction Temperature Range		T_J, T_{stg}	−65 to +150	°C

1. $I_C = 1.8$ A, L = 20 mH, P.R.F. = 10 Hz, $V_{CC} = 10$ V, $R_{BE} = 100 \Omega$.

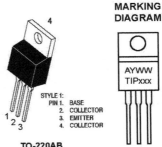

**MARKING
DIAGRAM**

AYWW
TIPxxx

STYLE 1:
PIN 1. BASE
2. COLLECTOR
3. EMITTER
4. COLLECTOR

**TO-220AB
CASE 221A-09
STYLE 1**

xxx = Specific Device Code:
31, 31A, 31B, 31C, 32, 32A, 32B, 32C
A = Assembly Location
Y = Year
WW = Work Week

ORDERING INFORMATION

See detailed ordering and shipping information in the package dimensions section on page 6 of this data sheet.

Publication Order Number:
TIP31A/D

TIP31, TIP31A, TIP31B, TIP31C, (NPN), TIP32, TIP32A, TIP32B, TIP32C, (PNP)

THERMAL CHARACTERISTICS

Characteristic	Symbol	Max	Unit
Thermal Resistance, Junction to Ambient	$R_{\theta JA}$	62.5	°C/W
Thermal Resistance, Junction to Case	$R_{\theta JC}$	3.125	°C/W

ELECTRICAL CHARACTERISTICS (T_C = 25°C unless otherwise noted)

Characteristic		Symbol	Min	Max	Unit
OFF CHARACTERISTICS					
Collector-Emitter Sustaining Voltage (Note 2) (I_C = 30 mAdc, I_B = 0)	TIP31, TIP32 TIP31A, TIP32A TIP31B, TIP32B TIP31C, TIP32C	$V_{CEO(sus)}$	40 60 80 100	- - - -	Vdc
Collector Cutoff Current (V_{CE} = 30 Vdc, I_B = 0)	TIP31, TIP32, TIP31A, TIP32A	I_{CEO}	-	0.3	mAdc
(V_{CE} = 60 Vdc, I_B = 0)	TIP31B, TIP31C, TIP32B, TIP32C		-	0.3	
Collector Cutoff Current (V_{CE} = 40 Vdc, V_{EB} = 0) (V_{CE} = 60 Vdc, V_{EB} = 0) (V_{CE} = 80 Vdc, V_{EB} = 0) (V_{CE} = 100 Vdc, V_{EB} = 0)	TIP31, TIP32 TIP31A, TIP32A TIP31B, TIP32B TIP31C, TIP32C	I_{CES}	- - - -	200 200 200 200	μAdc
Emitter Cutoff Current (V_{BE} = 5.0 Vdc, I_C = 0)		I_{EBO}	-	1.0	mAdc
ON CHARACTERISTICS (Note 2)					
DC Current Gain (I_C = 1.0 Adc, V_{CE} = 4.0 Vdc) (I_C = 3.0 Adc, V_{CE} = 4.0 Vdc)		h_{FE}	25 10	- 50	-
Collector-Emitter Saturation Voltage (I_C = 3.0 Adc, I_B = 375 mAdc)		$V_{CE(sat)}$	-	1.2	Vdc
Base-Emitter On Voltage (I_C = 3.0 Adc, V_{CE} = 4.0 Vdc)		$V_{BE(on)}$	-	1.8	Vdc
DYNAMIC CHARACTERISTICS					
Current-Gain - Bandwidth Product (I_C = 500 mAdc, V_{CE} = 10 Vdc, f_{test} = 1.0 MHz)		f_T	3.0	-	MHz
Small-Signal Current Gain (I_C = 0.5 Adc, V_{CE} = 10 Vdc, f = 1.0 kHz)		h_{fe}	20	-	-

2. Pulse Test: Pulse Width ≤ 300 μs, Duty Cycle ≤ 2.0%.

References

[1] D. A. Neamen. *Electronic Circuit Analysis and Design*. 2nd Edition. McGraw-Hill: New York, NY, 2001, pp. 19-20.

[2] M. H. Rashid. *Power Electronics Circuits, Devices, and Applications*. 2nd Edition. Prentice Hall: Englewood Cliffs, New Jersey, 1993, p. 39.

[3] P. Horowitz and W. Hill. *The Art of Electronics*. 2nd Edition. Cambridge University Press: New York, NY, 1989, p. 57.

[4] P. Horowitz and W. Hill. *The Art of Electronics*. 2nd Edition. Cambridge University Press: New York, NY, 1989, pp. 595-596.

[5] Fermi National Accelerator Laboratory. "Fermilab's Chain of Accelerators." http://www-bd.fnal.gov/public/proton.html#CR.

[6] P. T. Krein. *Elements of Power Electronics*. Oxford University Press: New York, NY, 1998, Chapter 11.

[7] A. I. Pressman. *Switching Power Supply Design*. 2nd Edition. McGraw-Hill: New York, NY, 1998, p. 28.

[8] W. H. Hayt, Jr. and J. E. Kemmerly. *Engineering Circuit Analysis*. McGraw-Hill: New York, NY, 1978, p. 333.

[9] N. Mohan, T. M. Undeland, and W. P. Robbins. *Power Electronics: Converters, Applications and Design*. 2nd Edition. John Wiley & Sons: New York, 1995, p. 27.

[10] S. Franco. *Design with Operational Amplifiers and Analog Integrated Circuits*. 3rd Edition. McGraw-Hill: New York, NY, 2002, pp. 276-278.

[11] R. W. Erickson and D. Maksimovic. *Fundamentals of Power Electronics*. 2nd Edition. Kluwer Academic Publishers: Norwell, MA, 1999, p. 68.

[12] P. T. Krein. *Elements of Power Electronics*. Oxford University Press: New York, NY, 1998, pp. 358-360.

[13] A. I. Pressman. *Switching Power Supply Design*. 2nd Edition. McGraw-Hill: New York, NY, 1998, Chapter 11.

[14] P. Horowitz and W. Hill. *The Art of Electronics*. 2nd Edition. Cambridge University Press: New York, NY, 1989, pp. 595-596.

[15] A. S. Sedra and K. C. Smith. *Microelectronic Circuits*. 5th Edition. Oxford University Press: New York, NY, 2004, p. 1245.

[16] H. Taub and D. Schilling. *Digital Integrated Electronics*. McGraw-Hill: New York, NY, 1977, pp. 9-10.

[17] ExtremeTech. "Suppression Components: Tubes, Diodes, and Circuits." http://www.extremetech.com/article2/0,1558,1155238,00.asp. (date accessed: December 18, 2004)

[18] ExtremeTech. "Suppression Components: Metal Oxide Varistors." http://www.extremetech.com/article2/0,1558,1155237,00.asp. (date accessed: December 18, 2004)

[19] P. W. Traynham and P. Bellew. *Using Thermally Protected MOVs in TVSs or Power Supply Applications*. Littlefuse Inc.: Melbourne, FL. http://www.littelfuse.com/data/Articles/TMOVPCIMPaper.pdf

Index